Practical Management of Tunneling with Tunnel Boring Machines

This book covers the management of mechanized tunneling with examples from global projects. It starts with an introduction to mechanized tunneling including management of job organization, planning job sites, portals, or launching boxes in mountains/open fields and urban areas. The management of the transport with belt conveyors, locomotives, and multi-service vehicles is explained with numerical examples. Cost management and basic parameters governing tunneling costs in different countries are discussed. Risk management in mechanized tunneling projects is also explained.

Features:

- Offers the practical issues with setting up a job site, the cost, and logistic issues related to tunneling.
- Reviews cost management and basic parameters governing tunneling costs in different countries.
- Covers treatment of spoil management plan and the management of contaminated ground.
- Explores key points on the logistics and the management of the consumables.
- Provides the latest international case studies of specific companies.

This book is aimed at professionals and researchers in tunneling, civil and mining engineering, and geology.

Practical Management of Tunneling with Tunnel Boring Machines

Nuh Bilgin and Sinan Acun

CRC Press
Taylor & Francis Group
Boca Raton London New York

CRC Press is an imprint of the
Taylor & Francis Group, an **informa** business

Cover photograph representing the assembly of the Robbins XRE-TBM in the chamber within the Gerede Tunnel, Turkey. With the kind permission of Timur Koloğlu, from Kolin Construction Company

First edition published 2024
by CRC Press
6000 Broken Sound Parkway NW, Suite 300, Boca Raton, FL 33487–2742

and by CRC Press
4 Park Square, Milton Park, Abingdon, Oxon, OX14 4RN

CRC Press is an imprint of Taylor & Francis Group, LLC

ISBN: 978-1-032-41622-9 (hbk)
ISBN: 978-1-032-41623-6 (pbk)
ISBN: 978-1-003-35897-8 (ebk)

DOI: 10.1201/9781003358978

Typeset in Times
by Newgen Publishing UK

Dedication

This book is dedicated to our lovely wives
Ayfer Bilgin and Dilek Acun
and our beloved children
Damlanur Bilgin, Derin and Deren Acun

Contents

Preface

The amount of tunneling with tunnel boring machines is increasing tremendously in the world. Unfortunately, some projects don't end on the scheduled time and with a planned budget, since mechanized tunneling is a complex process, needing careful coordination between each unit. In other words, good project management is the key to the success of the project. There are excellent books published on mechanized tunneling. However, it is scarce to find books on the shelf of a bookstore on the management of tunneling with TBMs. This book, which is intended to fill the gap in this respect, is a result of several years of experience of the authors and an intensive literature survey. The main topics covering the book are as follows: contractual practice, site set up, tunnel transport, cost management, logistics management of the consumables, job organization, segment production, management of the contaminated ground during and after TBM drives, risk management in mechanized tunneling and insurance issues, and management of tunnel fire risks during TBM drives. Each topic will be treated with real-life examples. In a few cases, numerical examples will also be given as to how to dimension a portal in the open field.

This book is the third one in this series. The first one was *Mechanical Excavation in Mining and Civil Industries*, and the second was *TBM Excavation in Difficult Ground Conditions: Case Studies from Turkey*. And we believe that this book will be a supplementary contribution to the success of TBM tunneling.

Acknowledgments

The acknowledgments are due to Dr. Yalçın Eyigün the General Director of AYGEM of the Turkish Ministry of Transport and Infrastructure, to Murat Cebeci and Mustafa Öztürk, the project directors of Gayrettepe-Istanbul Airport and Halkalı-İstanbul Airport Metro Projects, for sharing their immense experiences during two symposiums organized by the Turkish Tunnelling Society. The authors of this book benefited to some extent from their presentations made in the symposiums. Our thanks are also to Timur Koloğlu who gave us the permission to use the unique photograph of TBM (Robbins XRE). The machine was assembled using Onsite First Time Assembly (OFTA) in the underground launch chamber assembly in Gerede Tunnel.

The contents of this book were discussed at some of the World Tunnelling Conferences organized by ITA, and some of the data has been published in different technical journals such as *Tunnelling and Underground Space Technology*. However, the topics in this book include more data and it has been analyzed more comprehensively. We are grateful to the organizers of the World Tunnel Congress and cited journal authorities.

About the Authors

Nuh Bilgin graduated from the Mining Engineering Department of Istanbul Technical University. He obtained his PhD at Newcastle upon Tyne University, UK, in 1977 on "Rock cutting mechanics of high strength rocks". He was appointed as a full-time professor in 1989 at İstanbul Technical University (ITU). He worked one year at Colorado School of Mines, USA, in 1994 as a visiting professor and scholar of Fulbright, and one year at the University of Witwatersrand, SA, in 1989 as a visiting professor.

He was formerly the Head of the Mining Engineering Department, Istanbul Technical University, and Senator of the school representing in the Senate of the university. However, at the moment he is working as a consultant for Mining and Tunneling Companies. He is currently the Chairman of the Turkish Tunnelling Society. He published more than 150 papers on Tunneling and Mining, including four books.

He was the coordinator of one of the most prestigious international projects carried out at ITU. The project was on the development of mining methods and mining machinery in the Bismarck Sea for the production of metallic ores. Nautilus Minerals, from Australia, sponsored the project.

Sinan Acun graduated from the Geology Department of Kocaeli in 2004. Since then, he has worked as chief geologist for ASM Engineering Company; site engineer at Garanti Koza – Alarko Joint Venture in 4. Levent – Ayazaga Metro Project; TBM tunnel chief engineer in Gülermak – Doğuş Joint Venture in Otogar-Bağcılar Metro Project; TBM tunnel shift engineer in Avrasya Joint Venture (Astaldi-Makyol-Gülermak) for Kadikoy – Kartal Metro Project; TBM tunnel section chief in Gülermak Heavy Industry in Kargı Hydropower Project; tunnel manager in Gülermak – Kolin-Kalyon Joint Venture for Mecidiyeköy-Mahmutbey Metro; TBM tunnels group manager in Kolin-Kalyon-Cengiz Joint Venture for Gayrettepe – İstanbul New Airport Line Metro Project. He worked as tunnel expert for Kolin Construction Company in Uşak-Eşme Salihli infracture Works for the Ankara-İzmir High Speed Train Line Project. Currently, he is working as tunnel TBM manager in İÇTAŞ Construction Company for Sarıyer-Kilyos highway road project using an EPB-TBM of 13.75 diameter.

He is continuously attending World Tunnel Congresses and publishes papers. His paper, as co-author, "The factors affecting the performance of three different TBMs in complex geology in Istanbul" was selected as the best paper in WTC Bergen.

1 Introduction

1.1 GENERAL

In urban areas, a wide range of underground structures has been used to improve living conditions. Tunnels for metros, motorways, water supply, sewerage, electrical, and telephone cables are a priority. In the case of interurban links, long tunnels are justified by saving time and reducing costs with shorter journeys and less energy consumption. In the last 50 years, advancing technology in mechanized tunneling, using tunnel boring machines (TBMs), has made tunneling safer, is five or six times faster than conventional tunneling, and the length of tunnels excavated at once has become longer and longer. An example of the longest tunnel is the Seikan Tunnel 53,850 m long in Japan with a cross section of 74 m^2, which was the longest railway tunnel until 2016, with an undersea section, running between Honshu and Hokkaido. The undersea section measures 23.3 km. Seoul Subway Line 5 of 51,700 m in length is the longest metro line in the world. In Seikan Tunnel beneath the Tsugaru Strait, the use of a TBM was abandoned after less than 2 km owing to the difficult ground conditions. Nevertheless, tunneling continued with traditional tunneling methods on 28 September 1971. Thirty-four workers were killed during construction. The tunnel was opened on 13 March 1988, having cost a total of US\$7 billion to construct, almost 12 times the original budget, much of which was due to inflation over the years (https://en.wikipedia.org/wiki/Seikan_Tunnel). Railway twin tube tunnel, Gotthard Base Tunnel in the Central Swiss Alps, 57,104 m and 57,017 m in length, respectively, is the longest railway tunnel. It is also the world's longest transit tunnel. The Gotthard Base Tunnel was opened on 1 June 2016 and full service began on 11 December 2016. With a route length of 57.09 km, it is the world's longest railway and deepest traffic tunnel. Four Herrenknecht Gripper TBMs were used in the construction of the tunnel. The total excavation length realized by TBM was about 45 km for each tube (https://en.wikipedia.org/wiki/List_of_longest_tunnels).

Although TBMs are safer and faster than conventional tunneling, the proper selection of the machine needs expertise in this area. The main factors governing the choice of a right TBM for a described job depend on many factors, such as geology, geotechnical factors, abrasivity of the formations, groundwater, and the height of the overburden. Different types of TBMs are well described by Rostami (2016).

DOI: 10.1201/9781003358978-1

There are excellent books on the topic, such as on the basic working principles of TBMs, cutting mechanism, and performance analysis in urban areas and in the mountains (Guglielmetti et al. 2007; Maidl et al. 2011, Maidl et al. 2013; Bilgin et al. 2014, 2016). One should also remember that the success of the mechanical excavation with TBMs depends also on the management of the project. However, none of these books are devoted to the management of tunneling with TBMs. This incompleteness is intended to be fulfilled with the aid of the book in your hands. As Really (2000) mentioned:

How can we improve the management and procurement of complex underground programs? More effective and efficient management of large, public-works underground design and construction projects is an important need in an era of reduced funding and dramatically increasing public expectations. Management, design, procurement and construction of complex, urban underground projects – in particular, those contracts involving sophisticated equipment such as fully mechanized tunnel boring machines (TBMs) – are subject to many complex and interrelated variables. These include national and local politics, public policy, legal requirements, community involvement, media attention, and strict environmental compliance.

Based on the several years of experience of the authors and on a large literature survey, this book has 14 chapters. We believe that each of these chapters is a must, leading to the success of the TBM project.

REFERENCES

https://en.wikipedia.org/wiki/Seikan_Tunnel. Downloaded on 8 October 2023.

https://en.wikipedia.org/wiki/List_of_longest_tunnels. Downloaded on 8 October 2023.

Bilgin N., Çopur H., Balci C., 2014. *Mechanical Excavation in Mining and Civil Industries*, CRC Press, London, p. 380.

Bilgin, N., Copur, H., Balci, C., 2016. *TBM Excavation in Difficult Ground Conditions. Case studies from Turkey*, Ernst & Sohn, Berlin, Germany.

Guglielmetti, V., Grasso, P., Mahtab, A., Xu, S., 2007. *Mechanized Tunnelling in Urban Areas, Design Methodology and Construction Control*, CRC Press, Leiden, The Netherlands, p. 528.

Maidl, B., Herrenknecht, M., Maidl, U., Wehrmeyer, G., 2013. *Mechanized Shield Tunnelling 2012*, Ernst & Sohn, Berlin, p. 490.

Maidl, B., Schmid, L., Ritz, W., Herrenknecht, Sturge, M., 2011. *Hardrock Tunnel Boring Machines*, Wiley, Berlin, p. 356.

Reilly, J.J., 2000. The management process for complex underground and tunneling projects. *Tunnelling and Underground Space Technology*, 15 (1), pp. 31–44.

Rostami, J., 2016. Performance prediction of hard rock tunnel boring machines (TBMs) in difficult ground. *Tunnelling and Underground Space Technology*, 57, pp. 173–182.

2 Mechanized Tunneling with TBMs

2.1 INTRODUCTION

TBMs are used as an alternative to conventional tunneling methods. They have the advantages of high advance rates, limiting the disturbance to the surrounding area and producing a smooth tunnel wall without overbreak. They are suitable for the construction of tunnels in urban and high-traffic areas having less surface disturbance. They are designed to operate in various ground conditions as they are capable of digging soft ground like sand and hard rocks. They offer a continuous operation, and a better working environment compared to the drilling and blasting tunneling method. The number of laborers required is also reduced. The major disadvantage of mechanized tunneling is the considerable initial cost for TBM and also for the supporting accessories and equipment such as conveyor belts, slurry separating plant, and slurry pipelines. The mobilization time required is higher, and TBMs offer limited flexibility in complex geological conditions. The longer the tunnel, the less the relative cost of mechanical excavation with TBMs versus drill and blast methods.

The first TBM used was invented in 1863 and improved in 1875 by British Army officer Major Frederick Edward Blackett Beaumont. During the late 19th and early 20th centuries, the improvements in TBM design and building continued for railroads, subways, sewers, and water supply tunnels. Some interesting examples are a TBM with a bore diameter of 14.4 m manufactured by The Robbins Company for Canada's Niagara Tunnel Project. The machine was used to bore a hydroelectric tunnel beneath Niagara Falls. An earth pressure balance (EPB)-TBM known as Bertha with a bore diameter of 17.45 m was produced by Hitachi Zosen Corporation in 2013. It was delivered to Seattle, Washington, for its Highway 99 tunnel project. The machine began operating in July 2013, but stalled in December 2013 and required substantial repairs that halted the machine until January 2016. Bertha completed boring the tunnel on 4 April 2017. Two TBMs supplied by CREG excavated two tunnels for Kuala Lumpur's Rapid Transit with a boring diameter of 6.67 m in water-saturated sandy mudstone, schistose mudstone, highly weathered mudstone, as well as alluvium. It achieved a maximum advance rate of more than 345 m/month. A considerable large EPB-TBM of 15.62 m diameter with a total weight of 4,500 tons and total installed capacity of 18 MW was built by Herrenknecht AG and used in the Sparvo gallery of the Italian Motorway Pass A1. The energy consumption of this machine was about

DOI: 10.1201/9781003358978-2

62 million kWh/year. The 3.340 km double-deck tunnel connecting the European and Asian sides of Istanbul was excavated with a 13.7 m diameter Herrenknecht Mixshield Slurry TBM exclusively designed and equipped with the latest 483 mm disc cutters. TBM operations were completed in 479 calendar days, resulting in an average advance rate of 7.0 m/day. The maximum advance rate was realized in the marine sediment zone at 18.0 m/day. Furthermore, 440 disc cutters, 85 scrapers, and 475 brushes were replaced by TBM crew, and four times hyperbaric maintenance operations (total of 45 days) with specially trained divers (max. under 10.8 bar or 1.08 MPa for the first time in the world) were successfully performed https://en.wikipedia. org/wiki/Tunnel_boring_machine.

2.2 GEOLOGICAL AND GEOTECHNICAL INVESTIGATIONS FOR MECHANIZED TUNNELING/TBMS

The success of a tunneling project with TBM depends mainly on the correct selection of the machine on the experience of the contractor and on the basic management issues. However, the most important parameter leading to the success of the project is definitely understanding the geology and geotechnical parameters of the ground. Bearing in mind all this, this chapter will be devoted to the understanding of the importance of field studies, side investigation, the geotechnical data report (GDR), and geotechnical baseline report (GBR).

2.2.1 Field Studies, Side Investigation, the Geotechnical Data Report (GDR), and Geotechnical Baseline Report (GBR)

Side investigation is carried out to provide data to evaluate the feasibility of a tunneling project including cost, productivity, and scheduling of each stage of the project. Geological and hydrogeological conditions determine to a great extent the planning and budget necessary to complete the project. In fact, unexpected geological conditions may even stop a project or even make it unfeasible. Therefore, before starting the project, it is absolutely necessary to obtain as precisely as possible all geotechnical data of the geological formations, soil, or rock. First a desk study on available geological data is necessary, and thereafter a detailed boring program should be carried out. Therefore, boring data will permit us to extrapolate geological and geophysical surveying data, allowing us to predict in advance the possible approximate cost and time for the construction of the project. Field studies lead to the preparation of a GDR and GBR. The GDR contains only factual data and presents the results of field and laboratory data for the project without including an interpretation of this data. The GDR should contain the following information (Essex 2007a, 2007b): descriptions of the geological setting, descriptions of the site exploration program(s), logs of all borings, trenches, and other site investigations, descriptions/discussions of all field and laboratory test programs, results of all field and laboratory testing. The GBR is generally used for defining the baseline conditions for contractors to select methods, equipment, and risk analysis. It also serves as the basis for bid preparation and is used extensively in resolving disputes during construction.

2.2.2 Laboratory Testing Methods

This section describes in general physical and mechanical properties of rocks and soils relevant to TBM tunneling. Sample preparation and testing are important for obtaining significant data. The most widely used standards are the American Society for Testing Materials (ASTM, 2013) and International Society for Rock Mechanics (ISRM, 2007), providing guidance for performing the laboratory and in situ tests and also non-standardized recommended tests such as full- and small-scale linear or rotary rock cutting (Tables 2.1 and 2.2).

Determining the optimum amount of soil conditioning agents and the behavior of excavated soil and foaming agent mixture during excavation process plays an important role in the design and management of EPB-TBM tunneling. Although there are some means of defining optimum soil conditioning parameters, Torvane shear testing device is proved to provide an inexpensive and easy method to measure the shear strength of conditioned soils for soil condition optimization of EPB-TBM tunneling.

2.2.3 Laboratory Small- and Full-Scale Cutting Tests for the Performance Estimation of TBMs

Rock-cutting experiments are the best choice for performance prediction, since it is reliable and gives the possibility of defining basic specifications of TBMs and designing their cutterheads. A more popular small-scale rock-cutting rig (SLCM) was used at the University of Newcastle upon Tyne, England, for core cutting at 5 mm of the depth of cut, and specific energy measured was then used for predicting the performance of roadheaders (McFeat-Smith and Fowell 1977, 1979) by using a standard tungsten carbide cutting tool. A similar small-scale rock cutting was also instrumented at Istanbul Technical University (Bilgin and Shahriar 1987) as seen in Figure 2.1. By using this SLCM testing system, Balci and Bilgin (2007) correlated the specific energy obtained from the SLCM tests with the specific energy obtained from the full-scale linear cutting machine (FLCM) tests, which also enabled them to predict the performance of TBMs. The second version of the SLCM as seen in Figure 2.2 (also named portable liner rock-cutting rig/PLCM) is being developed at Istanbul Technical University (ITU; Çomaklı et al. 2021). It includes a small and stiff reaction frame on which the cutter and load-cell assembly are mounted. A block sample in size up to 10 × 15 × 20 cm is cast within a metal sample box with fast-curing concrete at a certain dip angle, parallel or perpendicular to the bedding planes to simulate the different cutting conditions on a rock deposit. A servo-controlled hydraulic cylinder moves the sample box through the cutter at a preset (constant) depth of cut, cutter (line) spacing, and constant velocity (30 mm/s). The depth of cut and line spacing of the cutter can be adjusted by a mechanical device and a hydraulic cylinder, respectively. A constant cross-section disc cutter having a diameter of 145 mm and a tip width of 4.7 mm is used throughout the cutting experiments. A triaxial dynamometer is used to measure/record the orthogonal force components. Structural load capacity of the PLCM is 50 kN for normal force and 20 kN for rolling force. The tests performed for verification of the portable linear cutting machine indicate that the performance and operational

TABLE 2.1

Rock Testing Methods and Standards for Mechanical Miners and TBMs

Test Name		ISRM (2007) Suggested Methods	ASTM Standards	Other Recommended Methods
MECHANICAL STRENGTH	UCS	1979	D 2938	
	BTS	1978	D 3967	
	Static Elastic Constants	1979	D 3148	
	Dynamic elastic constants	1978	D 2845	
	Triaxial	1983	D 2664	
	Direct shear	1974	D 5607	
	Point load	1985	D 5731	
	Vickers			Nilsen and Ozdemir (1999)
	Siever's J			Nilsen and Ozdemir (1999)
	Schmidt hammer	1978	D 5873	
TOUGHNESS/ BRITTLENESS	Punch penetration test			Nilsen and Ozdemir (1999)
	Fracture toughness	1988		Nilsen and Ozdemir (1999)
	Brittleness value (S20)			Zare and Bruland (2013)
	Siever's J value (SJ)			Zare and Bruland (2013)
	Abrasion value (AV)			Zare and Bruland (2013)
	Abrasion value cutter steel (AVS)			Zare and Bruland (2013)
ABRASIVENES	Cerchar		D 7625	CSM (1996)
	Schimazek			Schimazek and Knatz (1970)
	NTNU AVS			Zare and Bruland (2013)
ROCK CUTTING	Small-scale linear rock cutting			Fowell and McFeat (1976) ; Balci (2004)
	Full-scale linear rock cutting			CSM (1996) Eskikaya et al. (2000)
OTHER	Petrographic analysis	1978		Nilsen and Ozdemir (1999)
	Sound velocity (P and S waves)	1978	D 2845	
	Density	1979		
	Porosity	1979		

Source: Bilgin et al. (2014).

TABLE 2.2
Soil Testing Methods and Standards for TBM Selection

Test Name		AASHTO*	ASTM Standards	Other Recommended Methods
PHYSICAL AND MECHANICAL PROPERTIES	Cohesion	T 236	D 3080	
	Angle of internal friction	T 236	D 3080	
	Specific gravity	T 100	D 854	
	Water content	T 265	D4959	
	Atterberg limits (liquid limit, plastic limit, and plasticity)	T 89	D 4318	
	Standard penetration test	T 206	D 1586	
	Unconfined compressive strength of cohesive	T 208	D 2166	
	Triaxial strength	T 296, 297	D 2850, 4767	
	Modulus of elasticity			USACE EM 1110-1-1904
	Direct shear strength	T 236	D 3080	
PETROGRAFIC AND OTHER PROPERTIES	Grain size distribution (sieve and hydrometer analysis)	T 88	D 422, D 2487, D1140	
	Mineral contents (quartz content)			X-Ray Diffraction
	Clay mineralogy			X-Ray Diffraction
	pH		D 4972	
	Groundwater conditions			
	Sulfate content	T 290	D 4230	
	Chloride content	T 291	D 512	
	Abrasiveness (grain shape and hardness)		ASTM G75	Nilsen et al. (2006)
	Swell potential of clays	T 256	D 4546	
	Collapse potential of clays		D 5333	
	Permeability for granular soil	T 215	D 2434 D 5084	
	Permeability for all soil			
	Possible existence of boulders and cobbles: type, amount, sizes, strength, and abrasivity			

Source: Bilgin et al. (2014).

Note: *American Association of State Highway and Transportation Officials.

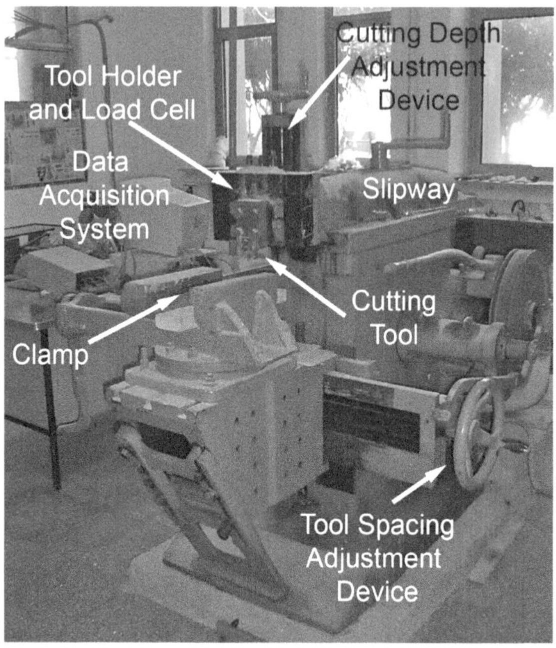

FIGURE 2.1 Early design of the small-scale rock-cutting rig. (Bilgin et al. 2014.)

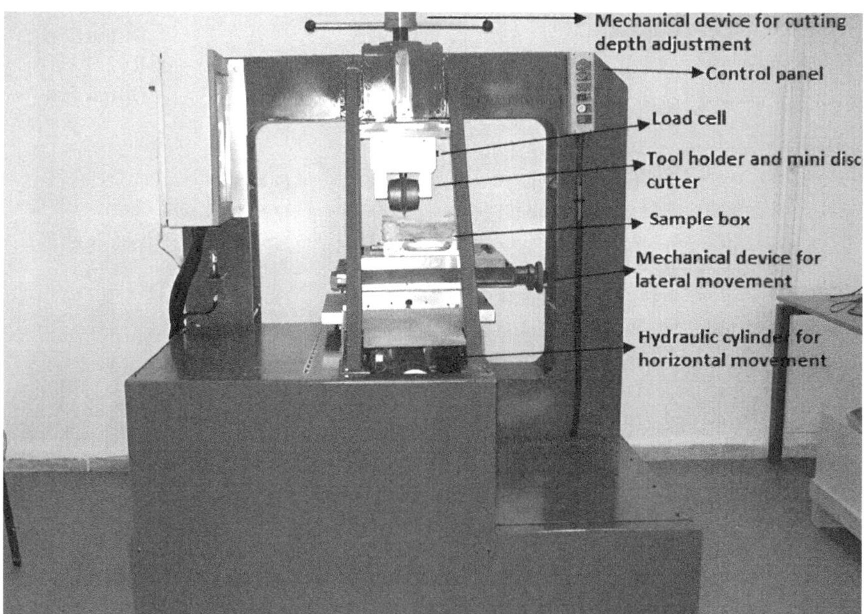

FIGURE 2.2 The second version of small-scale rock-cutting rig (also named portable liner rock-cutting rig developed at ITU). (From the archive of the author, N. Bilgin.)

FIGURE 2.3   Full-scale linear rock-cutting machine in ITU. Photograph by Bilgin.

parameters of the two identical EPB-TBMs, such as net cutting rate, thrust force, cutterhead torque, and power, are reliably predicted theoretically (deterministically) by using the portable linear cutting machine test results.

The full-scale linear cutting tests measure full-scale forces acting on a real-life cutter of any type (single disk, conical, radial, and so on) while cutting a block of rock sample cast in a sample box, the rig developed in ITU is seen in Figure 2.3. Full-scale testing minimizes the uncertainties of scaling and any unusual rock-cutting behavior not reflected in its physical properties. The results of this test can be used as input for selection, designing, and predicting the costs and performance (excavation/production/cutting rate) of TBMs mechanical miners, including for feasibility purposes. This test, along with deterministic computer simulation, is accepted as the most reliable and economical method for these purposes. Although the FLCM testing method used in different research institutes and universities has some disadvantages such as requiring experienced personnel, large blocks of rock samples and longer time periods for testing. However it is more reliable method for machine performance estimation. The currently used FLCMs are designed to keep the depth of cut constant and to measure normal, cutting, and side forces such as the FLCM used in Colorado School of Mines (Ozdemir 1990), Istanbul Technical University (Bilgin et al. 1999, 2005, 2014), Korea Institute of Construction Technology and Seoul National University (Chang et al. 2009; Cho et al. 2013), CSIRO Earth Science and Resource Engineering (Shao et al. 2014), and Beijing University of Technology (Zhao et al., 2015) and other research institutes in China. The details of the cutting rig developed in ITU are well defined by Bilgin et al. (2014). The measured and predicted values of TBM performances using this experimental cutting rig in different case studies are proved reliable and realistic (Bilgin et al.1999; Balci 2009; Bilgin 2017). A typical example of how to use the cutting test results will be given in the section dealing with the performance prediction of TBMs.

2.3 HARD ROCK AND MIX FACE TBMS, TYPICAL EXAMPLES FROM THE RECENT PROJECTS AND LESSONS LEARNED

There are basically three types of hard rock TBMs: gripper/open type, single shield TBM, and double shield TBM. The basic criteria for selecting each type of TBM and their working principles will be explained in this section. Typical examples of each type of TBM will be given with the geology, geotechnical characteristics of the ground, TBM performance parameters, and problems encountered during the excavation with working pie charts. We believe that all this information will clarify the readers leading on the correct selection of TBM and the performance prediction of these fascinating machines of the century, and contribute to the management of tunneling with TBMs.

2.3.1 GRIPPER/OPEN-TYPE TBMS, WORKING PRINCIPLES, A CASE STUDY, LESSONS LEARNED

A schematic view of a typical gripper/open-type TBM is seen in Figure 2.4.

2.3.1.1 Working Principles of Gripper/Open-Type TBMs

The open/gripper TBM is the most traditional among the hard rock TBMs and is the fastest and most effective at tunneling in competent/massive hard rock formations. During the boring operation, in the rear part of the cutting head, the gripper is firmly anchored to the tunnel walls by two large pads while the front part of the TBM is

FIGURE 2.4   Gripper/open-type TBM. Courtesy of Herrenknecht.

extended by the action of propel thrust cylinders. In this section, working principles will be explained by referring to the numbers in Figure 2.4. When tunneling with a gripper TBM, thrust cylinders (5) brace themselves against the gripper shoes (6) and push the cutterhead (1) against the tunnel face. The cutterhead is equipped with disc cutters and buckets (2). Due to the rolling movement of the disc cutters, chips are broken out of the rock. Buckets take up the excavated material and convey it to the muck chutes and to a belt conveyor (4). Gripper shoes (6) brace the gripper laterally against the tunnel wall using hydraulic cylinders. Roof bolting unit (5) drills boreholes for the rock bolts to ensure safety directly behind the roof shield (3). After the completion of a stroke, tunneling is interrupted and the gripping unit is moved forward with the aid of walking shoes (7). Herrenknecht Gripper TBMs have excavated and secured more than 85 km of the main tubes in Gothard base tunnels, while breaking world records (https://en.wikipedia.org/wiki/Tunnel_boring_machine).

2.3.1.2 A Case Study, Baltalimani Tunnel, Some Problems Which May Occur During Tunneling with Gripper-Type TBMs

The first TBM used in Istanbul was a gripper TBM (Figure 2.5). It worked in heavily fractured Trakya formation and moderately fractured Büyükada formation. The experience was a nightmare for the contractor, and the machine was withdrawn

FIGURE 2.5 The first TBM used in Istanbul. (Bilgin et al. 2016.)

TABLE 2.3
TBM Performance in Büyükada and Trakya Formation

Parameter	Büyükada Formation	Trakya Formation
Machine utilization (%)	28.5	7.2
Machine downtime (%)	71.50	92.6
Net cutting rate (m/h)	1.22	1.7
Progress rate (m/h)	0.35	0.13
Mean shift advance (m/shift)	3.15	1.24
Best shift advance (m/shift)	11.50	9.57
Lowest shift advance (m/shift)	0.60	0.2
Mean daily advance (m/day)	7.18	3.12
Best daily advance (m/day)	20	16.5
Lowest daily advance (cm/day)	0.22	0.5
Mean weekly advance (m/week)	43	21
Best weekly advance (m/week)	46	66
Lowest weekly advance (m/week)	9.95	1.9
Mean monthly advance (m/month)	197	84
Best monthly advance (m/month)	261.4	177.65
Lowest monthly advance (m/month)	56.2	17.33

Source: Bilgin et al. (2016).

and replaced with a shielded roadheader. The tunnel was opened in Büyükada and Trakya formations frequently cut by andesite and diabase dykes of 1–30 m thickness. Büyükada formation, of Upper Devonian Age, consists of micritic and nodular limestone and carbonate-rich shale. It is strongly folded, slightly jointed, and has a massive appearance. The joints are generally perpendicular to the bedding and perpendicular to the sub-vertical dip. Trakya formation consists of mudstone, shale-greywacke, and conglomerate units. It is closely jointed and strongly folded. RQD in Trakya formation is comparatively very low.

The geological cross section of the tunnel with a summary of the geotechnical data is seen in Figure 2.6 and some of the detailed TBM performance values are given in Table 2.3. As seen from this figure and table, the machine utilization is very low and is at an unacceptable level, like 7.5% in the Trakya Formation, having values changing from 5% to 20% and Q values from 0.01 to 1.5. However, this value is around 29% in the Büyükada formation, which is a relatively more competent rock with higher RQD and Q values. Machine utilization factor is a very important parameter which defines the efficiency of tunneling and is defined as the percentage of the time used just for excavation (excluding stoppage time) over a whole shift time (stoppages + excavation time). RQD is rock quality designation, a number decreasing with the fracturing of the rock mass, changing from 100% to 0% and it is determined from rock cores or volumetric joint count; it is the percentage of rock

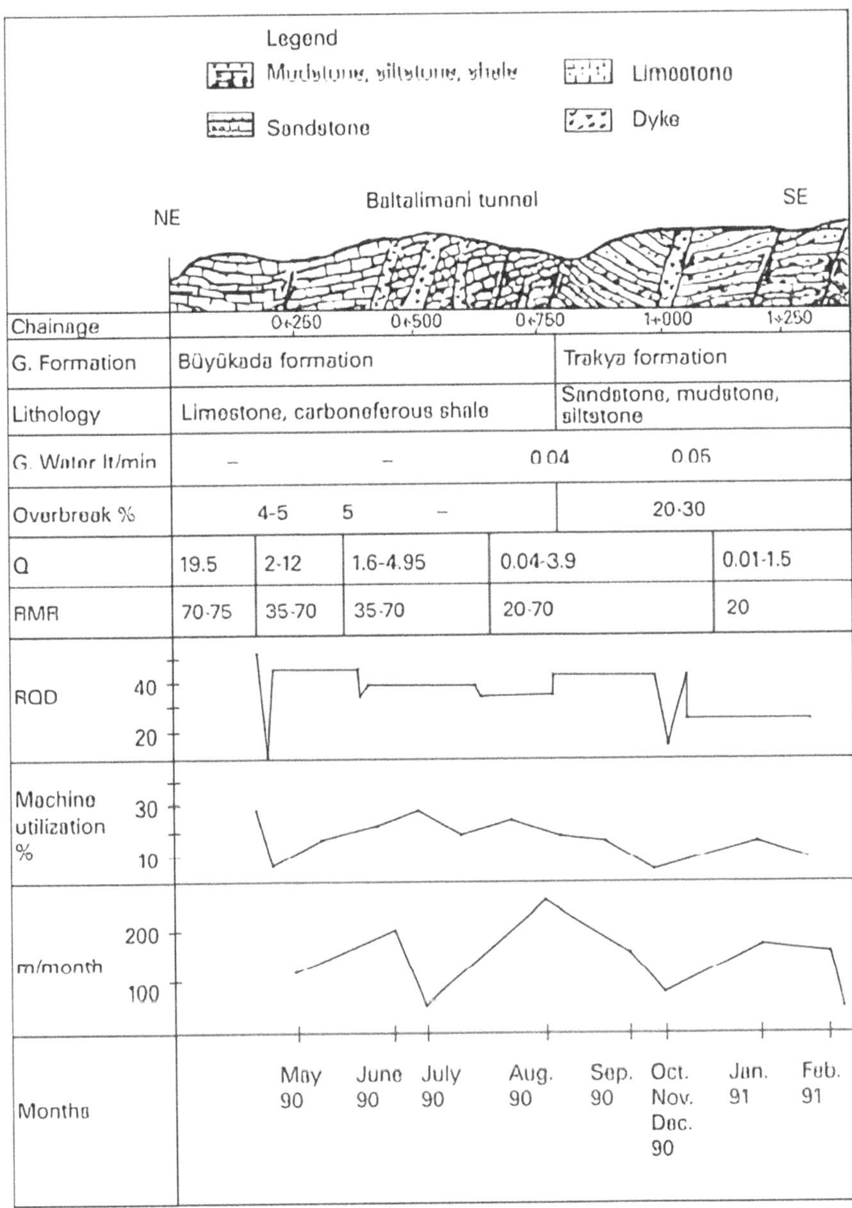

FIGURE 2.6 Geological cross section of Baltalimani Tunnel with TBM performance values. (Bilgin et al. 2016.)

cores having length less than 10 cm in 1 m of drill run. The rock mass rating (RMR) is a geomechanical classification system for rocks, which was developed by Z. T. Bieniawski between 1972 and 1973 (Bieniawski 1989). On the basis of RMR values for a given engineering structure, the rock mass is sorted into five classes: very good (RMR 100–81), good (80–61), fair (60–41), poor (40–21), and very poor (<20). The Q system for rock mass classification was developed by Barton et al. (1974). It expresses the quality of the rock mass in the so-called Q value, on which designs are based and to support recommendations for underground excavations. As seen in Table 2.3, low RQD, Q value, and RMR value are also reflected in mean shift advance, mean daily, weekly, and monthly advance rates. The mean shift advance rate in relatively competent rock is 2.5 times higher than in weak rock formations. One of the most important lessons that Turkish contractors learned from this case

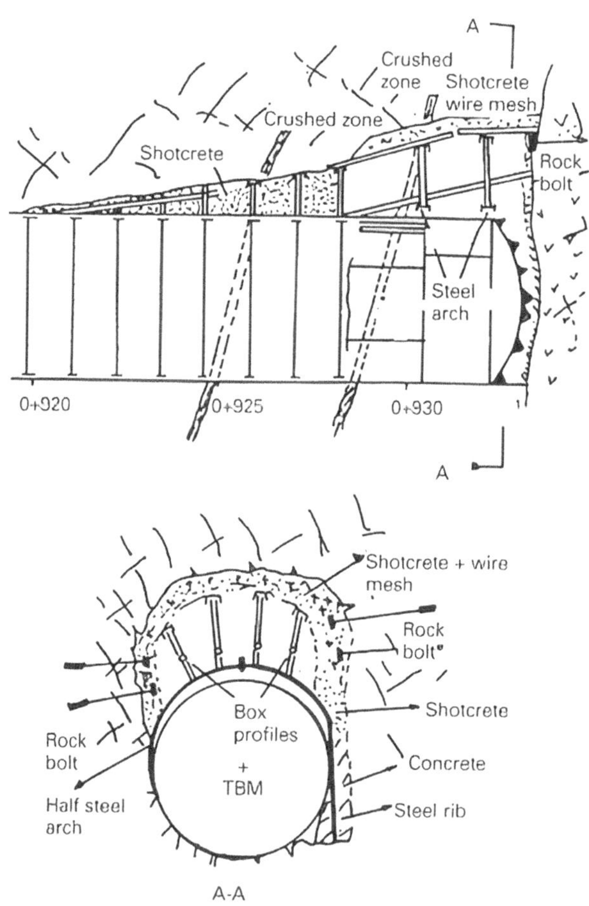

FIGURE 2.7 Collapse between chainage 0 + 920 and 0 + 930. (Bilgin et al. 2016.)

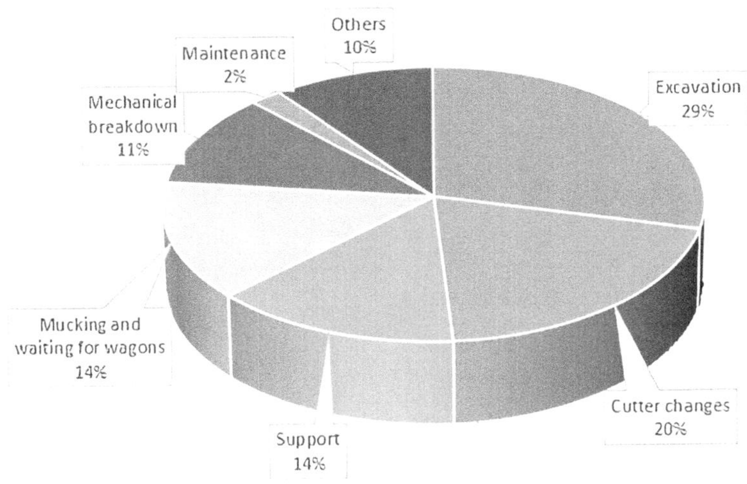

FIGURE 2.8 Working pie chart in Baltalimani Tunnel within chainages 0 + 250 and 0+500 km. (Bilgin et al. 2016.)

study was that a gripper TBM should never be used in rock formations having low RQD, Q value, and RMR values; otherwise, tunnel face collapses like happened in Baltalimani tunnel are inevitable as seen in Figure 2.7. The second important lesson emerging from work studies done between 0 + 250 and 0 + 500 km, as in Figure 2.8, is directly related to management issues. Waiting at a value of 14% is for mucking and waiting for wagons waiting for cutter changes is 20%. With a better job organization and selecting appropriate good quality disc cutters, these numbers could be reduced considerably.

2.3.2 Single Shield TBMs

A schematic view of a single shield TBM is seen in Figure 2.9.

2.3.2.1 Working Principles of a Single Shield TBM

During tunneling in low-strength rock mass, tunnel walls cannot react against the pushing of the gripper pads. In this case, one alternative to the gripper/open type TBM is a single shield TBM. The single shield TBM reacts against precast segment to advance forward, and due to this, installation of segments and excavation cannot be done simultaneously. Referring to the numbers in Figure 2.9, buckets (1) take up the excavated material and convey it to the muck ring and belt conveyors (2). Thrust cylinders (3) push the shield forwards from the previously built precast segments (4). Front shield (5) provides high safety for tunnel crew, especially in brittle rock. The annular gap (6) between the excavated ground and the outside of the tunnel lining is

FIGURE 2.9 A schematic view of a single shield TBM. Courtesy of Herrenknecht.

filled continuously with grout or pea gravel to minimize ground movement. Erector (7) serves to install segments during the ring building.

2.3.2.2 Case Study, Uluabat (Turkey) Tunnel, Some Problems Which May Occur During Tunneling with Single Shield TBMs

Tunnel excavation commenced in 2012 from the portal next to Uluabat Lake, using conventional excavation methods with roof bolts, shotcrete, wire mesh, and steel arches as the primary tunnel support. Tunneling operations were halted in November 2003 due to extreme roof deformations and floor heaves and it was decided to continue the project with a 5.05 m diameter EPB-TBM. Although the machine was an EPB-TBM, it worked most of the time in an open mode so the case may be an example of a single shield TBM. The geological cross section of the tunnel is seen in Figure 2.10 and some geotechnical data of the formations and operational parameters of TBM are given in Tables 2.4 and 2.5. The Uluabat Tunnel case study and the detailed TBM performance analysis presented clearly showed how the squeezing characteristic of the ground can impact the progress rate of a single shield EPB-TBM. Eighteen rescue galleries were constructed to free the trapped TBM. The days that were spent on TBM rescue operations were 192 (Bilgin and Algan 2012). An average daily advance rate of 8.6 m/day was achieved including all stoppages such as TBM standstills and hand mining. The best daily and weekly advance rates were found to be 28.8 m and 198.4 m, respectively. The best monthly advance rate was 583.2 m in February 2007. The learning points from this case study emerge from Figures 2.11 to 2.12. In the squeezing ground, the time spent on the installation of rings may go up to 43%, which is expected to be 25% in non-squeezing ground, and machine utilization time is highly related to Q values. Machine is trapped with Q values going up to 0.20 and machine utilization time goes up steadily to a level of 25% and levels off steadily.

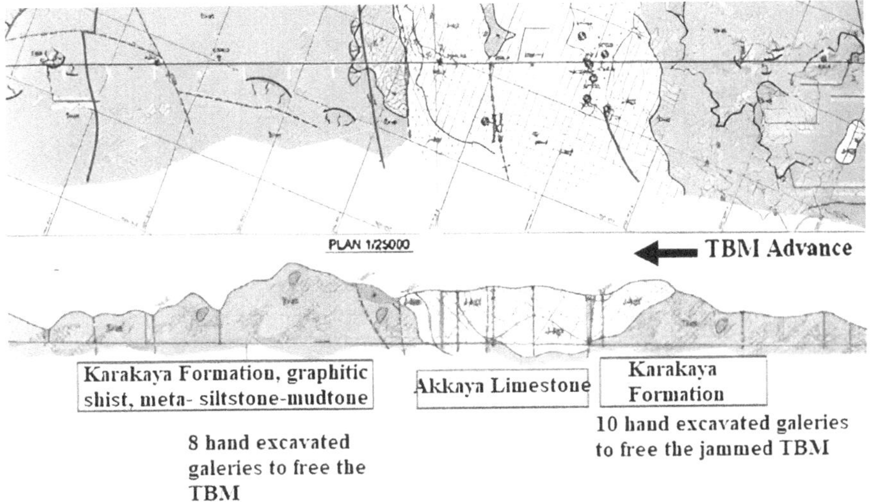

FIGURE 2.10 Geological cross section of Uluabat Tunnel. (Bilgin and Algan 2012.)

TABLE 2.4
Some Geotechnical Data in Uluabat Tunnel

Formation	UCS (MPa)	BTS (MPa)	E (GPa)	RMR
Karakaya	1–96	2.5–4	1.53–39.5	III–V
Akçakaya	24–125	3.5–5.4	9.1–126.6	III

Source: Bilgin and Algan (2012).

Note: UCS: Compressive strength; BTS: Brazilian tensile strength; E: Elastic modulus; RMR: Rock mass rating.

TABLE 2.5
Operational Parameters of the TBM Used in Uluabat

TBM Parameter	Value
TBM diameter	5.05 m
Number of discs	34
Disc diameter	17 inch
Number of scrapers	48
Maximum thrust force	29,000 kN
Nominal torque	2,048 kN at 6.25 rpm
Maximum power	2,100 kW
TBM weight	335 t

Source: From the archive of Bilgin.

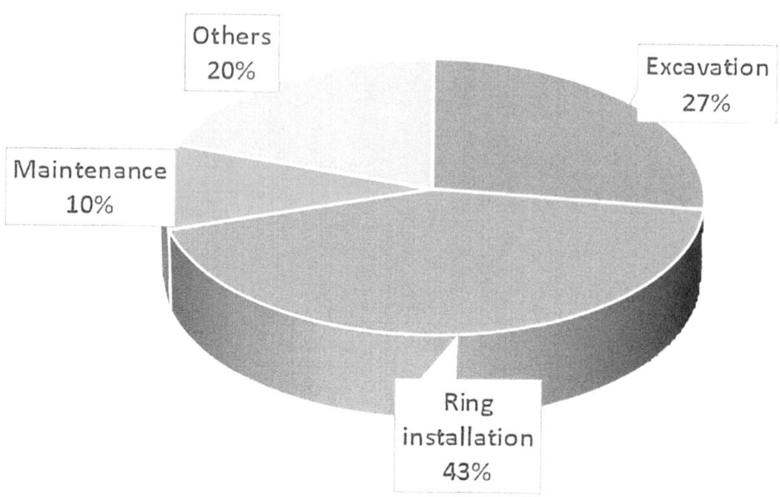

FIGURE 2.11   Pie chart from work study in Uluabat Tunnel. (From the archive of N. Bilgin.)

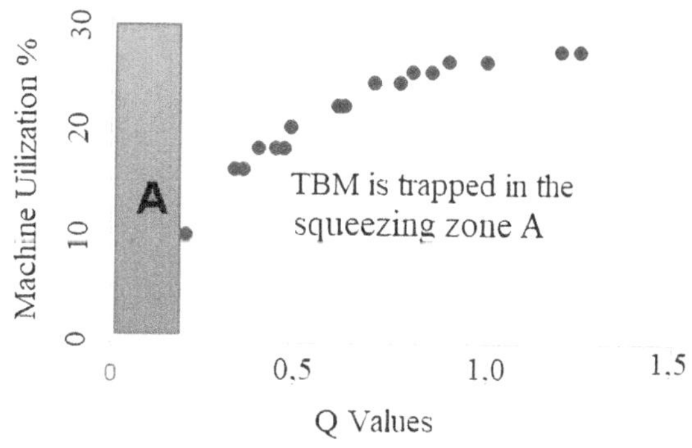

FIGURE 2.12   Variation of machine utilization time with Q values in a squeezing geological formation in Uluabat. (Barton and Bilgin 2016.)

2.3.2.3   Case Study on a Small Diameter Single Shield TBM, Tarabya (İstanbul) Tunnel, Some Problems Associated

Tarabya Tunnel with an excavation diameter of 2.9 m is part of a sewerage project, which was planned to clean the sea pollution Tunnel in Istinye – Tarabya and Büyükdere bays around Istanbul Bosphorus. Rock formations in the tunnel and some mechanical properties of rock formations, general characteristics of TBM (Herrenknecht) used in, and its performance chainage 981 and 7,700 m are summarized in Tables 2.6, 2.7, and 2.8. The results of the work studies carried out between these chainages are given in Figures 2.13 and 2.14. As seen from these figures, the average machine utilization time

TABLE 2.6
Rock Formations in the Tunnel and Some Mechanical Properties

Rock Formation (% of the total)	Compressive Strength (MPa)	Tensile Strength (MPa)	Elastic Modulus (MPa × 10³)
Limestone 65%	44–81	4–5	9–15
Shale 17%	55–59	2–4	9–10
Sandstone-siltstone 12%	59	–	–
Dykes 1%	32–40	3	6–7

Source: Bilgin et al. (2005).

TABLE 2.7
General Characteristics of TBM (Herrenknecht) Used in Tarabya

TBM excavation diameter	2,915 mm
TBM shield diameter	2,870 mm
Disc number	20
Disc diameter	305 mm
Max. rotational speed	16 rpm
Max. torque applied	725 kNm
Number of thrust cylinders	6
Max. thrust force	3,750 kN
Max. power	620 kW

Source: Bilgin et al. (2005).

TABLE 2.8
Overall TBM Performance

The best daily advance	24.75 m
The mean daily advance	11.50 m
The best monthly advance	531 m
The mean monthly advance	378 m

Source: Bilgin et al. (2005).

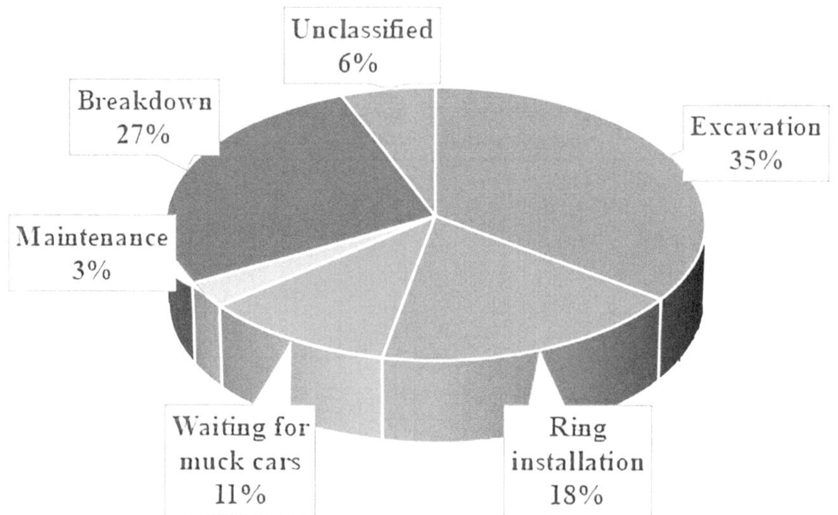

FIGURE 2.13   Pie chart of the work study in Tarabya Tunnel. (Bilgin et al. 2005.)

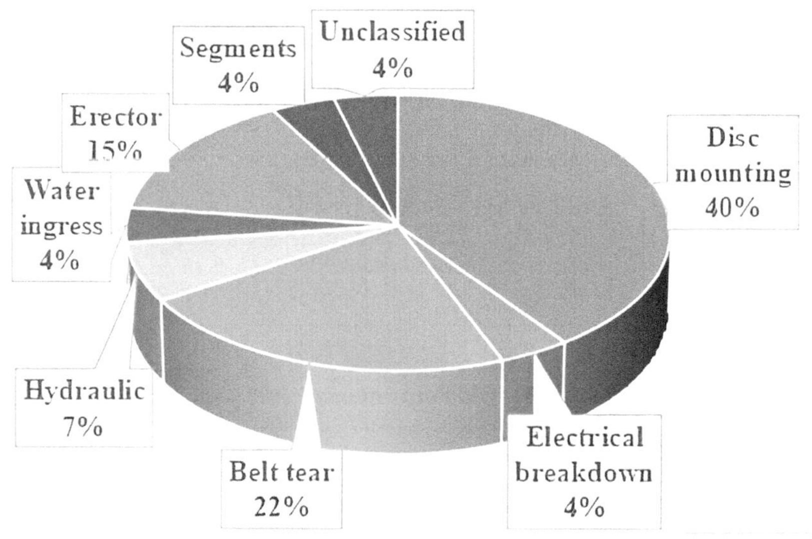

FIGURE 2.14   Pie chart of the breakdowns in Tarabya Tunnel. (Bilgin et al. 2005.)

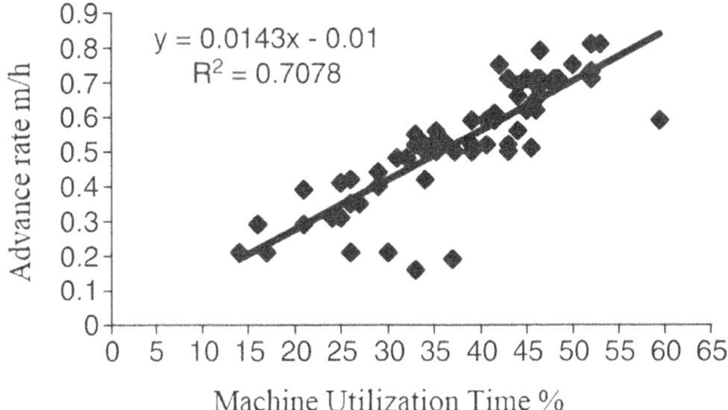

FIGURE 2.15 The variation of the machine utilization time with advance rate. (Bilgin et al. 2005.)

is 35% and the share of the breakdowns is 27% of shift time. However, it is interesting to note that the share of disc changing within the breakdown time is between 36% and 41%, which is quite high compared to other applications. The machine utilization time was found to change between 15% and 55% in general. Machine utilization time is one of the most important factors in determining the machine advance rate. As seen in Figure 2.15, an increase in machine utilization time from 15% to 55% increases the machine advance rate from 0.2 m/h to 0.8 m/h. Machine utilization time is defined as the ratio of the net cutting time of the machine to the total working time, i.e., shift time.

2.3.3 DOUBLE SHIELD TBMS, WORKING PRINCIPLES, A CASE STUDY, LESSONS LEARNED

A schematic view of a single shield TBM is seen in Figure 2.16.

2.3.3.1 Working Principles of a Double Shield TBM

Double shield TBMs have the combined method of gripper and single shield TBM. They are mainly used for tunnel projects in changing rock formations. They achieve a high advance rate in stable rock by working in continuous tunneling mode. Double shield TBM propulsion is achieved also by using a gripper device (4), but in difference from the open TBM, the front and rear parts of the TBM are completely shielded (1, 2, 3). The TBM installs a concrete lining ring using segment erectors (6). TBM advances with the aid of pushing cylinders (5). Precast segments perfectly protect the tunnel from the surrounding ground as the cutterhead excavates, allowing excellent advance rates. The numbers in the parentheses are from Figure 2.16.

FIGURE 2.16   A schematic view of a double shield TBM. Courtesy of Herrenknecht.

TABLE 2.9
Physical and Mechanical Properties of the Rock Formations

Rock	Chainage (km)	UCS (MPa)	Es (MPa)	C (MPa)	Φ^0	RQD (%)
Chalky limestone	30 + 000 34 + 500	10–44	8,468–15,126	4.51–5.17	27.7^0–30.8^0	15–70
Clayey limestone	34 + 500 36 + 300	4–7	–	–	–	25–45
Claystone Marn	36 + 300 46 + 470	2–13.5	963–5,692	3.1–3.95	14–29.9	15–85

Source: Ilci et al. (2013).

Note: UCS: compressive strength; Es: static elastic modulus; C: Cohesion; Φ^0: Internal friction angle; RQD: Rock quality designation.

2.3.3.2   A Case Study, Suruç (Turkey) Tunnel, Some Problems Which May Occur During Tunneling with a Double Shield TBMs

Suruç Tunnel, with a length of 17,185 m, is the longest irrigation tunnel ever being excavated in Turkey. It is planned to irrigate the Eastern Anatolian part of Turkey. The main rock formation in the area is Gaziantep formation of Eocene–Oligocene aged karstic chalky clayey limestone. Clay bands sometimes have 50–75 cm thickness. Ground water was encountered in the majority of drill holes. Mechanical properties of the rock formations are given in Table 2.9. A double shield hard rock TBM provided by Seli was used in Suruç Tunnel. The main characteristics of TBM are as follows: TBM diameter, 7.88 m; cutting head power, 4,500 kW; maximum torque, 4,000 kNm; maximum thrust, 20,000 kN; weight, 1,080 t. Weakly working hours were 168, and around 26 people worked per shift. Although the mean advance rate was one of the highest than the ones obtained with mechanized tunneling in Turkey by the time of boring Suruç Tunnel, the clogging effect of clayey formation on

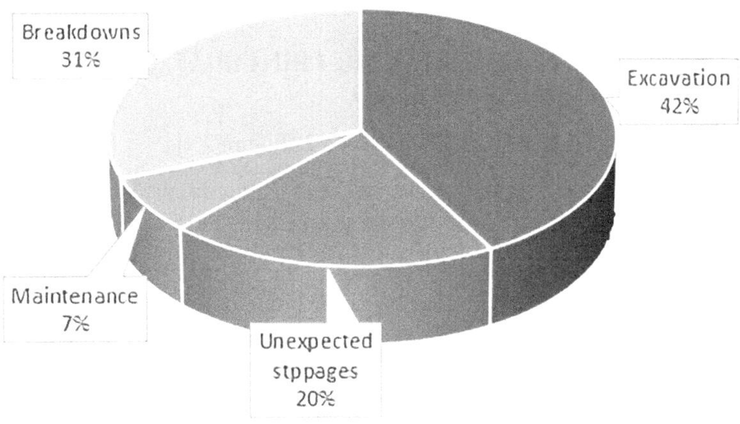

FIGURE 2.17 The work distribution for last 3 km. (Ilci et al. 2013.)

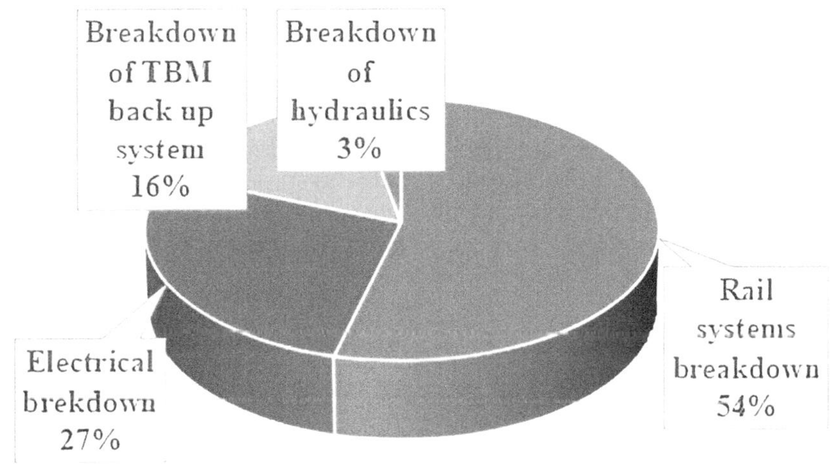

FIGURE 2.18 The percentage of breakdown within all stoppages. (Ilci et al. 2013.)

the cutterhead and on muck transport systems decreased the excavation efficiency. Work distribution for the last 3 km is given in Figure 2.17. Breakdowns, as seen in Figure 2.18, may be classified as follows: 54% rail systems breakdown, 27% electrical breakdown, 16% breakdown of TBM backup systems, and 3% breakdown of TBM hydraulics. One lesson learned from this case study was that although in favorable conditions the advance rate of double shields is high, management issues on muck transport may obscure the success of this machine. Rail system breakdowns of 54% of the total breakdown are definitely too high. This may not be seen in muck transport by belt conveyors.

2.4 CROSSOVER TBMS, WORKING PRINCIPLES, TWO CASE STUDIES, LESSONS LEARNED

A schematic view of a crossover TBM is seen in Figure 2.19.

2.4.1 WORKING PRINCIPLES OF CROSSOVER TBM

The machine has multi-speed gearboxes to enable the machine to advance through blocky, fractured, or mixed ground at high torque and low speed. Standing for a Crossover (X) between Rock (R) and EPB (E), the XRE has been designed and deployed by Robbins on multiple projects that feature sections of both hard rock and soft ground in the tunnel alignment. Using features of both EPB and single shield hard rock machines, the XRE has successfully been used on mixed ground projects in Australia, Turkey, Mexico, India, and also in some more countries. Readers are advised to have more technical information on this revolutionary machine to read a paper by Harding and Alpagut (2020).

2.4.2 CASE STUDIES ON CROSSOVER TBM, GEREDE TUNNEL

The Gerede Tunnel is the longest and most problematic water transmission tunnel ever bored in Turkey, with a length of 31.6 km and an overburden of 600 m. High-pressure water ingress and collapses have resulted in many delays. Although construction began in 2012, and was intended to be completed in 2014, the breakthrough could be realized in December 2018. The tunnel passed through permeable Tertiary sedimentary and volcanic sedimentary rocks, and Jurassic–Cretaceous limestone

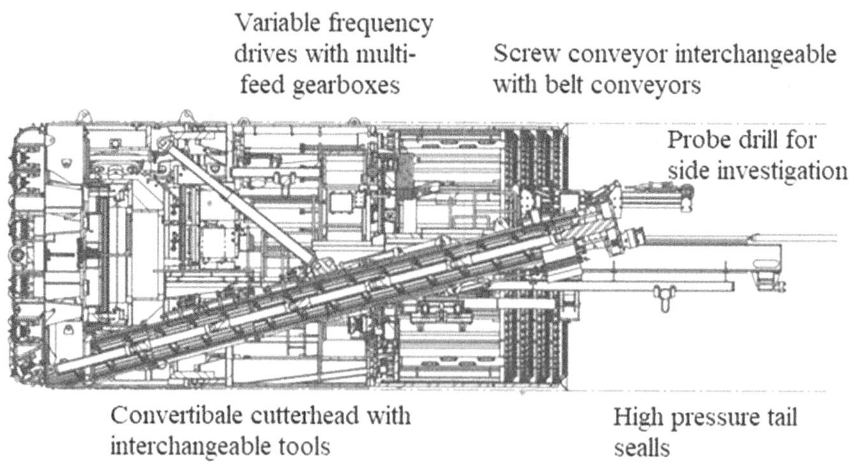

FIGURE 2.19 A schematic view of a crossover TBM. Courtesy of Robbins.

(Apaydın et al. 2019). Out of three standard double shield TBMs originally supplied to bore the tunnel, two became irretrievably stuck following massive inflows of mud and debris. There were some discrepancies between the predicted in the project and the actual geology. The length of the limestone having severe problems in terms of groundwater, which is estimated to be 750 m along the tunnel in the project, actually increased to 3.5 km. Therefore, the water problem encountered in this part was much more than expected. Plastic clays in the limestone, which were never mentioned in the project report, also compressed the double shield TBM and delayed finishing the tunnel. In most of the fault zone, breccia and loose agglomerates with high water pressures decreased the advance rate and even caused long stoppages several times. In 2016, a hybrid-type "crossover" machine was launched to excavate the final 9 km of tunnel. TBM was assembled and launched more than 7 km from the tunnel portal and successfully passed 48 fault zones as well as hydrostatic pressures up to 26 bar. Despite the challenges, crews were able to achieve a best month of 484 m with a 285 m monthly average throughout tunneling. December 2018 was the breakthrough of a 5.5 m diameter hybrid-type crossover machine which took its place in the literature with several items to be remembered (Harding and Alpagut 2020).

2.4.3 A Second Example to Crossover TBM, Salihli Eşme Tunnel

The 13.77 m XRE TBM was launched in March 2021. The machine bored 3.05 km on the Esme-Salihli Railway Tunnel as part of the Ankara–İzmir High Speed Railway Project, in a formation consisting of gneiss, mudstone, and sandstone gneiss. A team made up of employees from Kolin Construction, Turkish State Railways (TCDD), and Robbins witnessed the breakthrough of the world's fastest TBM over 13 m in diameter. The machine set world records three times over, beating its own records in May and June with a set of records over the summer, including a best day of 32.4 m, a best week of 178.2 m, and a best month of 721.8 m.

2.5 SOFT GROUND TBMS, TYPICAL EXAMPLES FROM THE RECENT PROJECTS AND LESSONS LEARNED

For tunneling projects in soft-ground geology with small overburden, generally carried out in urban areas, the TBM has to control accurately the pressure in the tunnel face, which must be higher than the one inside the boring machine.

2.5.1 Earth Pressure Balance Tunnel Boring Machines (EPB-TBM), Working Principles, and Lessons Learned

The EPB-TBM evacuates the muck through a screw conveyor. The pressure in the front of the machine is controlled by regulating the rotational speed of the screw conveyor. The TBM can work with or without pressure in the front in stable ground. EPB-TBMs are normally used for excavation of fine sand, silt, and clay having low permeability. They are not very effective in soils having fine material of less than 10% and water heads over 4 bar. They can also be used for hard rock excavations if their cutterheads and muck transportation units are suitable, of which these types can be considered as mixshields. A schematic view of an EPB-TBM is given in Figure 2.20.

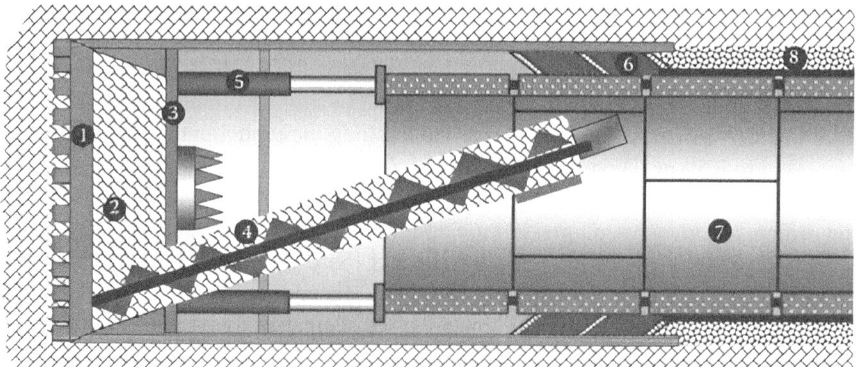

1) Cutterhead, 2) Pressure chamber, 3) Pressure wall, 4) Screw conveyor
4) Screw conveyor, 5) Thrust cylinders, 6) Tail sailant, 7) Segment,
8) Annular grout

FIGURE 2.20   A schematic view of an EPB-TBM. (Bilgin et al. 2014.)

2.5.1.1   A Case Study on an EPB-TBM from Halkalı–New Istanbul Airport

New Istanbul airport is being constructed over an area of 76.5 million square meters, 35 km away from the city center upon an old open cast coal field. It is planned to offer flights to more than 350 destinations. Twenty EPB-TBM worked within Gayrettepe–Halkalı–New Istanbul Airport Lines. There are few lakes left after mining activities along the metro line which cause tremendous difficulties during tunnel excavations. This section of the chapter concerns about the performance of an EPB-TBM under a lake between the chainages 32 + 340 and 32 + 640 km. The experience gained during tunnel excavation served better to understand the effect of EPB pressure on TBM thrust, penetration, torque, and specific energy values including the geotechnical parameters of the ground (Aslanbaş and Bilgin, 2020). Foam and chemicals used in complex geology, sticky ground, or when passing underwater are one of the key factors for a successful tunneling operation and should be determined after careful laboratory tests carried out on samples collected from relevant areas. A typical sample seen in Figure 2.21 and Table 2.10 gives the geotechnical characteristics of clayey sand silty sand. The following data on soil conditioning belongs to the chainage 32 + 400/32 + 700 km under the lake (Figure 2.22). The values were obtained after a wide range of laboratory tests carried out by chemical companies.

During the tunnel excavation, the muck samples were also sent to the laboratories to check the validity of the chemicals decided to be used. The mean value of FER (foam expansion ratio) was 40 with minimum and maximum values of 27 and 57. The mean value of FIR (foam injection ratio) was 8 with minimum and maximum values of 5 and 10. The mean ratio of foam/water ratio was 1.62. The foam used per ring (1.5) m was 43.4 L. Foam with an "Anti-clogging agent" was used to cope with clogging behavior of the sticky ground. Polymer-added foam and chemicals used in complex geology, sticky ground, or when passing underwater

FIGURE 2.21 General view of muck sample from clayey sand. (Aslanbaş and Bilgin 2020.)

TABLE 2.10
Geotechnical Characteristics of Clayey Sand Silty Sand

Geotechnical Parameter	Clayey Sand tda2–1	Silty Sand tda3
SPT (N30)	40	24
Permeability (Lugeon)	0.8	28
Liquid limit (%)	42	34
Plastic limit (%)	19	17
Plasticity index	23	16
Water content (%)	20	23
Density (kN/m^3)	22	22
Compressive strength (MPa)	2.8	2.7
Shear strength (kPa)	45	24
Internal friction (angle degree)	5–24	33

Source: Aslanbaş and Bilgin (2020).

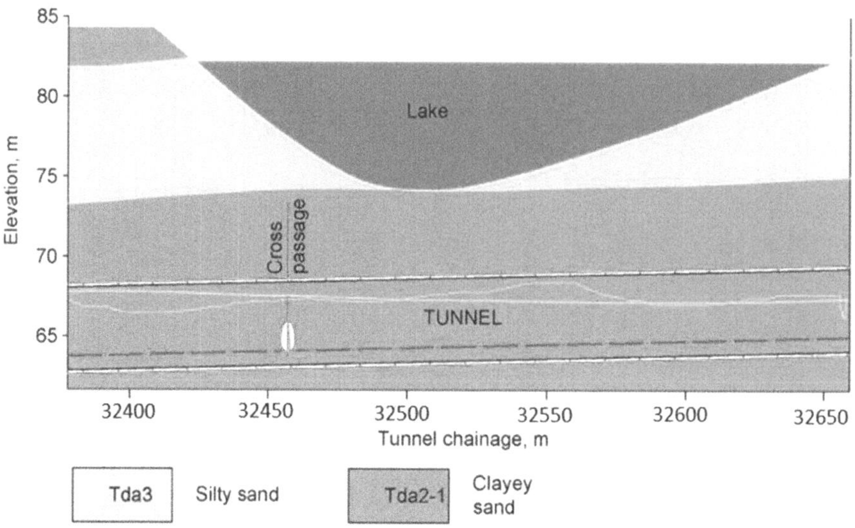

FIGURE 2.22 The geological cross section of the tunnel. (Aslanbaş and Bilgin 2020.)

are one of the key factors for a successful tunneling operation and should be determined after careful laboratory tests carried out on samples collected from relevant areas. A mean daily advance rate of 26.8 m/day was obtained for this section of the project with an EPB-TBM which basic design parameters are given in Table 2.11. Based on Figure 2.23, which shows the work study carried out in the tunnel under the lake, the applicant learned that management points such as logistic problems, compulsory waiting, and breakdowns limited the time spent for the excavation to 24%.

2.5.2 SLURRY TUNNEL BORING MACHINES, WORKING PRINCIPLES, AND LESSONS LEARNED

The slurry shield TBM is suitable for excavating soils with high water content. This TBM is equipped with a slurry system which controls the pressure in the excavation face by injecting pressurized slurry into the cutter chamber where the slurry is mixed with the excavated material. The mixture is pumped out of the tunnel to a separation and recirculation plant. A schematic view of a slurry TBM is seen in Figure 2.24.

2.5.2.1 A Case Study on a Slurry TBM, Avrasya Tunnel

The Eurasia Tunnel is a road tunnel in Istanbul, Turkey, crossing underneath the Bosphorus Strait. The tunnel was officially opened on 20 December 2016. The 3,340 m tunnel was bored and constructed with a 13.7 m diameter Herrenknecht Mixshield Slurry TBM exclusively designed and equipped with 483 mm disc cutters all atmospherically changeable and have individual wearing sensor system, and 192

TABLE 2.11
Design Parameter of EPB-TBM

Parameter	Value
TBM excavation diameter (m)	6.57
Shield diameter (m)	6.54; 6.53; 6.52
Shield length (m)	8.61
Shield thickness (cm)	5
Number of single discs	32 + 8
Disc diameter (inch)	17
Number of scrapers	72
Opening ratio (%)	25
Maximum rotational speed (rpm)	4.5
Maximum torque (breakout) (kNm)	6,635
Maximum thrust (kN)	51,090
Maximum power (kW)	3×400

Source: Aslanbaş and Bilgin (2020).

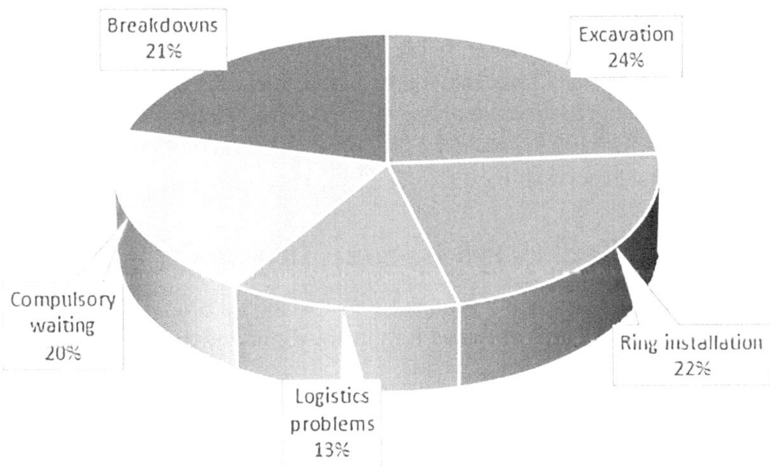

FIGURE 2.23 The results of work study carried out in the tunnel beneath the lake. (Aslanbaş and Bilgin 2020.)

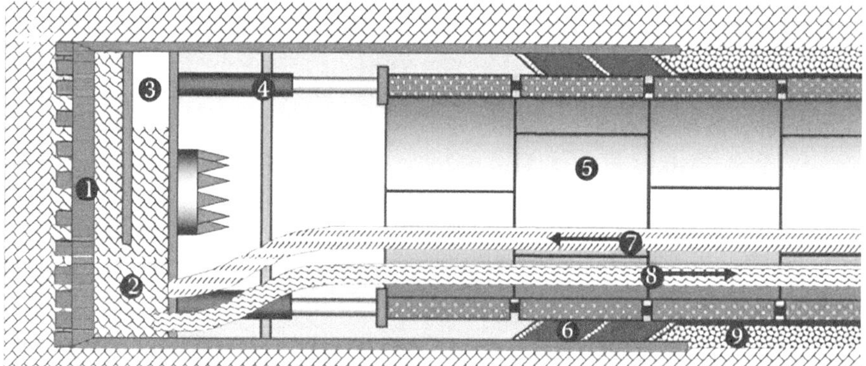

1) Cutterhead, 2) Bentonite slurry/soil, 3) Air bubbles, 4) Thrust cylinders
5)Segments, 6)Tail sailant, 7) Bentonite slurry feed, 8) Bentonite slurry/soil return,
9) Annular grout

FIGURE 2.24 A schematic view of a slurry TBM. (Bilgin et al. 2014.)

cutting knives were also used. Slurry treatment plant had an installed power of 3.5 MW and had a capacity of 2,800 m^3/h. The utilized TBM ranks first in the world with its 33 kW/m^2 cutterhead power, and second with its 12.0 bar face pressure design. TBM operation was completed within 479 calendar days, resulting in an average advance rate of 7.0 m/day by three crews 24/7. The maximum advance rate was realized in the marine sediment zone at 18.0 m/day. In Trakya formation, 28 dyke zones were excavated with an average frequency of 90 m and thicknesses varied between 1 and 120 m. Furthermore, 440 disc cutters, 85 scrapers, and 475 brushes were replaced by TBM crew and hyperbaric maintenance four times in a total of 45 days, with specially trained divers being successfully done.

2.5.3 The Comparison of EPB and Slurry TBMs

After all, one may ask when and why an EPB or a slurry TBM will be more efficient for this tunnel? Table 2.12 is prepared for the answer to this question.

2.5.4 Variable Density TBMs

As Babler et al. (2018) emphasize, the variable density TBM can be operated as classic slurry TBM with an air bubble system to control the face pressure and in a full EPB mode. The change between the modes can be done gradually under permanent and full control of the tunnel face pressure and without any need of chamber interventions. This machine can also be operated using a high density in the excavation chamber that would be too dense for classic slurry operation but that would be too fluid for a classic EPB operation. For all operation modes of the variable density TBM, the muck is extracted from the excavation chamber via a screw conveyor. For the first and second phase of the Kuala Lumpur Klank Vally project, Herrenknecht

TABLE 2.12
The Comparison of EPB and Slurry TBMs

Criterion	EPB-TBM	Slurry TBM
Torque requirement	Requires more torque	Requires lesser torque
Handling boulders	Boulders endanger screw conveyor	Handles easily using stone crusher units
Applicable to geological formations	Best applicable to clayey soils under water table	Best applicable to sandy soils under water table
Muck haulage	Muck haulage as it is	Hydraulic transportation
Muck separation plant	No need	Requires resulting extra cost and space
Face support	Conditioned muck with additives	Water plus bentonite slurry
Face pressure limit	4–4.5 bar	8.9 bar

Source: Bilgin et al. (2014).

delivered a total of ten variable density TBMs. The TBMs bored through the Kenny Hill formation in EPB mode and through the Kuala Lumpur Limestone and Granite formations in slurry mode and high-density slurry mode (Duhme and Brabant 2019). For further case studies the readers are advised to read the references cited for this topic.

2.6 PERFORMANCE PREDICTION OF HARD ROCK TBMS

One of the most important parameters in the success of a mechanized tunneling depends on how TBM will behave in a given geology. In other words, what will be the performance of the machine (thrust, torque, penetration, specific energy, and machine utilization factor)? Recently there have been excellent papers on the subject (Salimi et al. 2022; Farrokh 2020). However, the readers are advised to follow the references given below in order to understand the scientific development of the subject.

Several TBM performance prediction models with few on the prediction of machine utilization time were realized in the past. Full-scale rock-cutting tests coupled with geological and geotechnical parameters were the basis for predicting TBM performance. The works carried out at Newcastle upon Tyne University, Colorado School of Mines in the 1970s, and recently at ITU and at Korea Institute of Construction Technology and Seoul National University and some institutes in China are typical examples of full-scale laboratory rock-cutting experiments. Roxborough (1969), Roxborough and Phillips (1975), Bilgin (1977), and McFeat-Smith and Fowell (1977, 1979) carried out experiments at Newcastle Upon Tyne University. Ozdemir (1990), Rostami and Ozdemir (1993), Rostami et al. (1994, 1996), Ozdemir and Nilsen (1993), and Copur (1999) did basic rock-cutting tests in Colorado School of Mines. Balci (2009) and Bilgin et al. (2010, 2014) carried out full-scale tests at ITU. Chang et al. (2009) did some detailed rock-cutting studies at the Korea Institute of Construction Technology and Seoul National University. The work carried out by Dollinger et al. (1998) is another example of using specially designed punch penetrating tests for estimating TBM performance. The Norwegian University of Science and Technology (NTNU) model developed at the Norwegian University of Science and Technology is a semi-theoretical model which defines first cutter forces and later TBM performance using geological parameters (Bruland 1998). Other prediction models have been developed in recent years to account for a wide range of rock properties and rock mass conditions. The QTBM (Barton 2000) is based on an expanded Q system which can be used for TBM performance estimation. Hassanpour et al. (2009a, 2009b, 2011) analysed the TBM performance in relation to the field penetration index previously. Gong and Zhao (2007, 2009) and Gong et al. (2007) introduced the boreability index and rock brittleness and investigated rock mass characteristics affecting TBM performance. Sapigni et al. (2002) and Ribacchi and Fazio (2005a, 2005b) analyzed the relationship between TBM performance and RMR. Khademi et al. (2010) and Yagiz (2008) created TBM performance prediction models taking account the influence of the rock mass parameters. Bieniawski et al. (2007, 2008) and Bieniawski and Grandori (2007) investigated TBM borability using rock mass classification systems. Factors affecting TBM penetration rates in different rock formations were also investigated by Nelson et al. (1983, 1985). Various models for estimation of the penetration rate of hard rock TBMs were summarized recently by Farrokh et al. (2012, 2013). The problems of TBM excavation in blocky ground were investigated by Delisio and Zhao (2013) and Delisio et al. (2013).

2.7 PERFORMANCE PREDICTION OF EPB-TBMS

Although there is a lot of research work done on the performance prediction of hard rock TBMs, as explained in Section 2.6, the work done on the performance prediction of EPB-TBMs is rare.

2.7.1 A STOCHASTIC METHOD FOR PREDICTING PERFORMANCE OF EPB-TBMs, WORKING IN SEMI-CLOSED/OPEN MODE

As Copur et al. (2014) emphasized, the rock excavation process is a stochastic nature, which includes uncertainties in terms of decision-makers. These authors summarized a new method for predicting the performance (instantaneous penetration rate, daily advance rate, thrust, power, torque, and specific energy) of EPB-TBMs by a stochastic method implemented into a deterministic prediction model requiring input from full-scale laboratory linear rock-cutting tests. Full-scale linear rock-cutting experiments using a disc cutter were performed on a block of limestone sample. Estimations were made using a commercially available Monte Carlo simulation program providing knowledge of probabilistic outcomes. Results of the suggested method were verified by measuring the field performance of an EPB-TBM excavating a massive rock mass of limestone in semi-closed mode. Results indicate that the stochastic estimates give an opportunity to see uncertainties and variation in the experimental parameters through the use of estimated probability density in the function.

2.7.2 METHODOLOGY USED FOR PREDICTING EPB-TBM PERFORMANCE IN HARD FRACTURED ROCK FORMATIONS IN COMPLEX GEOLOGY

In this work, a model based on past experiences obtained in Istanbul was used. The TBM data collected from Beykoz Utility Tunnel, Cayirbasi Water Tunnel, Kartal–Kadikoy Metro Tunnel, Pendik–Kaynarca Metro Tunnel and Uluabat Power Tunnel from Bursa, and current metro Uskdar–Umraniye–Sancaktepe–Cekmekoy projects were used to develop the model described in this section (Namlı and Bilgin 2017). Specific energy is the energy spent on excavating a unit volume of rock and it is one of the most important factors in determining the efficiency of rock excavation and it may be used to estimate net or instantaneous production/cutting rates of a mechanical excavator as given in Equation (2.1) (Rostami et al. 1994; Copur et al. 2001).

$$NPR = k.P / SE \qquad (2.1)$$

where NPR is net production rate in (m^3/h), k is the energy transfer ratio from the cutting head to the tunnel face (it is usually 0.8 for TBMs), P is power spent on excavating the rock for related SE, and SE is specific energy in (kWh/m^3). Power is directly related to the rotational speed of the cutterhead, to torque, hence the rolling force of the cutters. In other words, the geotechnical properties of the rock, type or cutters, and design of the cutterhead may be calculated using empirical equations given in Namlı and Bilgin (2017). Key points in this method of calculation are first to calculate specific energy with the aid of Figure 2.25, to calculate machine utilization time using Table 2.13, and to use proper correction numbers depending on EPB values. Readers may consult the numerical example of this performance calculation methodology, which is given in detail in Namlı and Bilgin (2017).

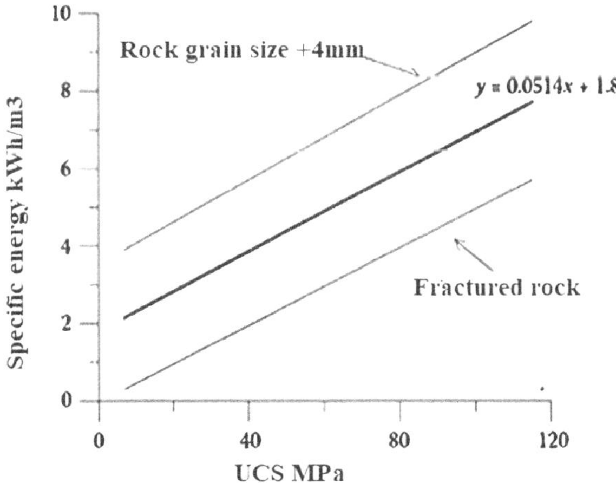

FIGURE 2.25 The variation of specific energy with uniaxial compressive strength.

TABLE 2.13
The Guidance to Estimate Machine Utilization Time

Stoppage Type		Stoppage Duration (% of one day shift time)
Adverse ground	Contact zones between geological formations	A few days to 1 week
	Dykes	A few days to 1 week
	Faults	A few days to 2 weeks
	Water	A few days to 1 week
TBM breakdown	New TBM (experienced crew)	2%–4%
	New TBM (inexperienced crew)	4%–6%
	Refurbished TBM (experienced crew)	4%–6%
	Refurbished TBM (inexperienced crew)	6%–8%
Cutter replacement	Quartz content of 0%–20%	5%
	Quartz content greater than 20%, weak and blocky ground	5%–10%
	Quartz content greater than 20%, hard rock	10%
Muck transportation	By train: Transportation distance of 0–3 km	7%
	By train: Transportation distance greater than 3 km	10%
	By belt conveyor: Transportation distance of 0–3 km	5%
	By belt conveyor: Transportation distance greater than 3 km	7%
Maintenance	Experienced contractor and crew	10%
	Moderately experienced contractor and crew	15%
Setting the segments		20%–25%
TBM mobilization at stations		2–3 weeks
Other stoppages		10%–15%
Machine utilization		22%–45%

Source: Namli and Bilgin (2017).

2.8   CONCLUDING REMARKS

TBMs are used as an alternative to conventional tunneling methods. They have the advantage of high advance rates, limiting the disturbance to the surrounding area. However, it is strictly advised to understand the geology and the behavior of the ground during tunneling with TBM. For this, first side investigation must be carried out to provide data to evaluate the feasibility of a tunneling project including cost, productivity, and scheduling of each stage of the project. Geological and hydrogeological conditions determine to a great extent the planning and budget necessary to complete the project. For hard rock formations, rock-cutting experiments are the best choice for performance prediction of the machine. It is reliable, and gives the possibility of defining basic specifications of TBMs and designing their cutterheads. There are basically three types of hard rock TBMs and three types of soft ground TBMs. Gripper/open type, single shield, and double shield are used for hard rock TBMs, EPB-TBMs, slurry TBMs, and high-density TBMs. The basic criteria for selecting each type of TBM and their working principles are explained in this chapter. Typical case studies of each type of TBM are given with the geology, geotechnical characteristics of the ground, TBM performance parameters, and problems encountered during the excavation with working pie charts. We believe that all this information will clarify the readers, lead to the correct selection of TBM and the performance prediction of these fascinating machines of the century, and contribute to the management of TBMs. One of the most important parameters in the success of a mechanized tunneling depends on how TBM will behave in a given geology. In other words, what will be the performance of the machine (thrust, torque, penetration, specific energy, and machine utilization factor)? Recently, there have been excellent papers on the subject. However, the readers are advised to follow the references given below in the chapter in order to understand the scientific development of the subject.

REFERENCES

https://en.wikipedia.org/wiki/Tunnel_boring_machine. Uploaded in September 2022.

Apaydin, A, Korkmaz, N, Ciftci, D., 2019. Water inflow into tunnels: Assessment of the Gerede water transmission tunnel (Turkey) with complex hydrogeology. *Quarterly Journal of Engineering Geology and Hydrogeology,* 52, pp. 346–349. https://doi.org/10.1144/qjegh2017-125

Aslanbaş, A., Bilgin, N., 2020. Excavating a metro tunnel under a lake with an EPB TBM used for Istanbul 3th Airport. *ITA-AITES World Tunnel Congress,* Kuala Lumpur.

ASTM. 2013. American Society for Testing and Materials. Standard test method for unconfined compressive strength of intact rock core specimens. Soil and Rock, Building Stones. *Annual Book of ASTM Standards,* Vols. 4.08 and 4.09.1828.

Babler, K., Battistoni, F., Burger, W., Aslanbaş, A., 2018. Variable density TBM – Combining two soft ground TBM technologies, *TBM Tunnel Business Magazine,* March 22.

Balci, C. 2004. *Comparison of small scale and full scale rock cutting tests to select mechanized excavation machines.* PhD Thesis, Istanbul Technical University, 269.

Balci, C., 2009. Correlation of rock cutting tests with field performance of a TBM in a highly fractured rock formation: A case study in Kozyatagi-Kadikoy metro tunnel, Turkey. *Tunnelling and Underground Space Technology,* 24, pp. 423–435. https://doi.org/10.1016/j.tust.2008.12.001

Balci, C., Bilgin, N., 2007. Correlative study of linear small and full-scale rock cutting tests to select mechanized excavation machines. *International Journal of Rock Mechanics and Mining Science,* 44, pp. 468–476. https://doi.org/10.1016/j.ijrmms.2006.09.001

Barton, N., 2000. *TBM Tunnelling in Jointed and Faulted Rock,* Balkema, Rotterdam.

Barton, N., Bilgin, N., 2016. Fast or slow progress with TBM in ideal or faulted condition, *Eurock 2016,* Cappodocia, Turkey.

Barton, N.R., Lien, R., Lunde, J., 1974. Engineering classification of rock masses for the design of tunnel support. *Rock Mechanics and Rock Engineering*, 6(4), pp. 189–236. doi:10.1007/BF01239496

Bieniawski, Z.T., 1989. *Engineering rock mass classifications: A complete manual for engineers and geologists in mining, civil, and petroleum engineering.* Wiley-Interscience, pp. 40–47.

Bieniawski, Z.T., Celada, B., Galera, J.M., 2007. Predicting TBM excavability – Part I. *Tunnels & Tunnelling International*, pp. 32–35 (September).

Bieniawski, Z.T., Celeda, B., Galeras, J.M., Tardaguila, I., 2008. New applications of the excavability index for selection of TBM types and predicting their performance, in *Proceedings of the World Tunnel Congress*, Akra, India, pp. 1618–1629.

Bieniawski, Z.T., Grandori, R., 2007. Predicting TBM excavability – Part II. *Tunnels & Tunnelling International*, 15–18 (December).

Bilgin, N., 1977. *Investigations into the Mechanical Cutting Characteristics of Some Medium and High Strength Rocks*, PhD Thesis. The University of Newcastle Upon Tyne, p. 332.

Bilgin, N., 2017. TBM performance prediction using laboratory cutting tests in very hard and abrasive rock formations. *3rd Conference on Tunnel Boring Machines in Difficult Ground Conditions*, 21–22, Wuhan, China.

Bilgin, N., Algan, M., 2012. The performance of a TBM in a squeezing ground at Uluabat, Turkey, *Tunnelling and Underground Space Technology*, 32, pp. 58–65. http://dx.doi.org/10.1016/j.tust.2012.05.004

Bilgin, N., Balci, C., Acaroglu, O., Tunçdemir, H., Eskikaya, S., Akgül, M., Algan, M., 1999. The performance prediction of a TBM in Tuzla–Dragos sewerage tunnel, in *The World Tunnel Congress'99, Oslo*, 31 May–3 June, Balkema, Rotterdam, pp. 817–822.

Bilgin, N., Balci, C., Tumac, D., Feridunoglu, C., Copur, H., 2010. Development of a portable rock cutting rig for rock cuttability determination, in *Proceedings of the European Rock Mechanics Symposium* (EUROCK 2010), J. Zhao, V. Labiouse, J.P.Duth, J.F. Mathier (eds), June 15–18, Lausanne-Switzerland, pp. 405–408.

Bilgin, N., Copur, H., Balci, C. 2014. *Mechanical Excavation in Mining and Civil Industries*, CRC Press, London.

Bilgin, N., Çopur, H., Balcı, C., 2016. *TBM Excavation in Difficult Ground Conditions, Case Studies from Turkey*, Erns and Sohn, Berlin.

Bilgin., N., Feridunoglu., C, Tumac, D., Cinar, M., Palakci, Y., Gunduz, O., Ozyol, L., 2005. The performance of a full face tunnel boring machine (TBM) in Tarabya (Istanbul), in *WTC, Underground Space Use: Analysis of the Past and Lessons for the Future*, Y. Erdem and S. Tülin (eds), Taylor & Francis Group, London.

Bilgin, N., Shahriar, K., 1987. *Development of an instrumented testing system and its application to TTK Amasra Coal Region, Istanbul Technical University; Mining Engineering Department.* Tubitak Project No: MAG674; (in Turkish).

Bruland, A., 1998. *Hard Rock Tunnel Boring Doctoral Dissertation*, Norwegian University of Science and Technology, Trondheim

Chang, S.-H., Choi, S.-W., Bae, G.-J., Jeon, S., 2009. Performance prediction of TBM disc cutting on granitic rock by the linear cutting test. *Tunnelling and Underground Space Technology*, 21, p. 271. https://doi.org/10.1016/j.tust.2005.12.131

Cho, J.W., Seakwon, J., Jeong, H., Chang, S.H., 2013. Evaluation of cutting efficiency during TBM disc cutter excavation within a Korean granitic rock using linear-cutting-machine testing and photogrammetric measurement, *Tunnelling and Underground Space Technology*, 35, pp. 37–54. doi:10.1016/j.tust.2012.08.006

Çomaklı., R., Balci, C., Copur, H., Tumac, D. 2021. Experimental studies using a new portable linear rock cutting machine and verification for disc cutters, *Tunnelling and Underground Space Technology*, 108, February, 103702. doi.org/10.1016/j.tust.2020.103702

Copur, H., 1999. *Theoretical and Experimental Studies of Rock Cutting with Drag Bits toward The Development of a Performance Prediction model for Roadheaders*, PhD Thesis. Colorado School of Mines. 361 p

Copur, H, Aydin, H, Bilgin, N, Balci, C, Tumac, D, Dayanc, C., 2014. A stochastic method for predicting performance of EPB TBMs, working in semi-closed/open mode, *ITA-AITES WTC, Brazil.*

Copur, H., Tuncdemir, H., Bilgin, N., Dincer, T., 2001. Specific energy as a criterion for used of rapid excavation systems in Turkish mines. *Transactions of the Institutions of Mining and Metallurgy Section A*, 110, pp. A149–A157.

CSM, 1996. Test and model descriptions for performance and cost description of mechanical excavators for mining, underground construction and micro tunnelling. *EMI of Mining Engineering Department*, 22.

Delisio, A., Zhao, J., 2013. Review of the TBM performance in blocky rocks with potential face stability issues. *World Tunnelling Congress*, 2013, Geneva, pp. 1179–1186.

Delisio, A., Zhao, J., Einstein, H.H., 2013. Analysis and prediction of TBM performance in blocky rock conditions at the Lötschberg Base Tunnel. *Tunneling and Underground Space Technology*, 33, pp. 131–142.

Dollinger, G.L., Handewith, J.H., Breeds, C.D., 1998. Use of punch tests for estimating TBM performance. *Tunnelling and Underground Space Technology*, 13(4), pp. 403–408.

Duhme., R, Brabant., J.D, 2019. Variable density TBMs as a solution for tunneling in difficult ground and flexible TBM usage. *The 2019 World Congress on Advances in Structural Engineering and Mechanics (ASEM19)*, Jeju Island, Korea, September 17–21.

Eskikaya, S., Bilgin, N., Ozdemir, L., et al. 2000. *Development of Rapid Excavation Technologies for the Turkish Mining and Tunneling Industries. NATO TU-Excavation SfS Programme Project Report*. Istanbul Technical University, Mining Eng. Dept., Sept., 172.

Essex, R.J. 2007a. Geotechnical baseline reports for construction – Second Edition, R.J. Essex (ed.), *North American Tunneling 2008 Proceedings*.

Essex, R.J. 2007b. Geotechnical baseline reports for construction suggested guidelines, R.J. Essex (ed.), *ASCE, American Society of Civil Engineers*, pp. 1–62.

Farrokh, E, 2020. A study of various models used in the estimation of advance rates for hard rock TBMs, *Tunnelling and Underground Space Technology,* 97, March 2020, 103219.

Farrokh, E., Rostami, J., Askilsrud, O.G., 2013. Down time analysis of hard rock TBN case histories. *World Tunnelling Congress 2013*, Geneva.

Farrokh, E., Rostami, J., Laughton, C., 2012. Study of various models for estimation of penetration rate of hard rock TBMs. *Tunnelling and Underground Space Technology,* 30, pp. 110–123.

Fowell, R.J., Mcfeat-Smith, I., 1976. Factors influencing the cutting performance of a selective tunnelling machine. *International Tunnelling*, 76, pp. 301–318.

Gong, Q.M., Zhao, J., 2007. Influence of rock brittleness on TBM penetration rate in Singapore granite. *Tunneling and Underground Space Technology*, 22, pp. 317–324.

Gong, Q.M., Zhao, J., 2009. Development of a rock mass characteristics model for TBM penetration rate prediction. *International Journal of Rock Mechanics and Mining Sciences*, 46, pp. 8–18.

Gong, Q.M., Zhao, J., Jiang, Y.S., 2007. In situ TBM penetration tests and rock mass boreability analysis in hard rock tunnels. *Tunneling and Underground Space Technology*, 22, pp. 303–316.

Harding, D, Alpgut, Y., 2020. Tunneling through 48 fault zones and high water pressures on Turkey's Gerede water transmission tunnel. *ITA-AITES World Tunnel Congress*, 12–15 May, Kuala Lumpur.

Hassanpour, J., Rostami, J., Khamehchiyan, M., Bruland, A., 2009a. Developing new equations for TBM performance prediction in carbonate-argillaceous rocks: A case history of Nowsood water conveyance tunnel. *Geomechanics and Geoengineering*, 4, pp. 287–297.

Hassanpour, J., Rostami, J., Khamehchiyan, M., Bruland, A., Tavakoli, H.R., 2009b. TBM performance analysis in pyroclastic rocks: A case history of Karaj water conveyance tunnel. *Rock Mechanics and Rock Engineering*, 43(4), pp. 425–445.

Hassanpour, J., Rostami, J., Zhao, J., 2011. A new hard rock TBM performance prediction model for project planning. *Tunneling and Underground Space Technology*, 26, pp. 595–603.

Ilci, N, Temel, M, Sezgin, S, Akpınar, T, Guarasio, S., Polat, C, Bilgin, N., 2013. Clogging and squeezing effect of marl-clayey limestone on the performance of a hard rock TBM in Suruc Tunnel, Turkey, in *WTC, 2013 Geneva, Underground – The Way to the Future!*, G. Anagnostou and H. Ehrbar (eds), Taylor & Francis Group, London.

ISRM. 2007. *The complete ISRM suggested methods for rock characterization, testing and monitoring 1974–2006.* In R. Ulusay and J.A. Hudson (eds), Compilation arranged by the ISRM Turkish National Group, Ankara, p. 628.

Khademi, H.J., Shahriar, K., Rezai, B., Rostami, J., 2010. Performance prediction of hard rock TBM using Rock Mass Rating (RMR) system. *Tunneling and Underground Space Technology*, 25, pp. 333–345.

McFeat-Smith, I., Fowell, R.J., 1977. Correlation of rock properties and the cutting performance of tunnelling machines. *Proceedings of the Conference on Rock Engineering*, Newcastle upon Tyne, pp. 581–602.

McFeat-Smith, I., Fowell, R.J., 1979. The selection and application of roadheaders for rock tunnelling. *Proceedings of the Rapid Excavation and Tunnelling Conference*, Atlanta, 1, pp. 261–279.

Namlı, M., Bilgin, N., 2017. A model to predict daily advance rates of EPB-TBMs in a complex geology in Istanbul, *Tunnelling and Underground Space Technolgy*, 62(1), pp. 43–52.

Nelson, P., O'Rourke, T.D., Kulhawy, F.H., 1983. Factors affecting TBM penetrationrates in sedimentary rocks. *Proceedings, 24th US Symposium on Rock Mechanics*, Texas A&M, College Station, TX, pp. 227–237.

Nelson, P.P., Ingraffea, A.R., O'Rouke, T.D., 1985. TBM performance prediction using rock fracture parameters. *International Journal of Rock Mechanics and Mining Sciences & Geomechanics Abstracts*, 22, pp. 189–192.

Nilsen, B., Dahl, F., Holzhäuser, J., Raleigh, P. 2006. SAT: NTNU's new soil abrasion test. *Tunnels & Tunneling International*, May 2006, pp. 43–45.

Nilsen, B., Ozdemir, L., 1999. Recommended laboratory rock testing for TBM projects, *American Underground Construction Association Journal*, 6, pp. 21–35.

Ozdemir, L., 1990. Recent developments in hard rock mechanical mining technologies. In *Proceedings of the 4th Canadian Symposium on Mining Automation*, September 16–18, Saskatoon, pp. 143–165.

Ozdemir, L., Nilsen, B., 1993. Hard rock tunnel boring prediction and field performance. In *Rapid Excavation and Tunneling Conference (RETC) Proceedings*, Boston, USA (Chapter 52).

Ribacchi, R., Fazio, A.L., 2005a. Influence of rock mass parameters on the performance of a TBM in a gneissic formation (varzo tunnel). *Rock Mechanics and Rock Engineering*, 38, pp. 105–127.

Ribacchi, R., Fazio, A.L., 2005b. Influence of rock mass parameters on the performance of a TBM in a gneissic formation (Varzo Tunnel). *Rock Mechanics and Rock Engineering*, 38, pp. 105–127.

Rostami, J., Ozdemir, L., 1993. A new model for performance prediction of hard rock TBMs, in *Proceedings of Rapid Excavation and Tunneling Conference*, L.D. Bowerman et al. (eds), Boston, MA, USA, pp. 793–809 (Chapter 50).

Rostami, J., Ozdemir, L., Bjorn, N., 1996. Comparison between CSM and NTH hard rock TBM performance prediction models. *ISDT*, Las Vegas.

Rostami, J., Ozdemir, L., Neil, D.M., 1994. Performance prediction: A key issue in mechanical hard rock mining. *Mining Engineering*, 11, pp. 1263–1267.

Roxborough, F.F., 1969. Rock cutting research for the design and operation of tunneling machines. *Tunnels and Tunnelling*, September, pp. 125–128.

Roxborough, F.F., Phillips, H.R., 1975. Rock excavation by disc cutter. *International Journal of Rock Mechanics and Mining Sciences & Geomechanics Abstracts*, 12, pp. 361–366.

Salimi, A., Rostami, J., Moormann, C., Hassanpour, J., et al., 2022. Introducing tree-based-regression models for prediction of hard rock TBM performance with consideration of rock type. *Rock Mechanics and Rock Engineering*, 55, pp. 4869–4891. https://doi.org/ 10.1007/s00603-022-02868-x

Sapigni, M., Berti, M., Bethaz, E., Busillo, A., Cardone, G., 2002. TBM performance estimation using rock mass classifications. *International Journal of Rock Mechanics and Mining Sciences*, 39, pp. 771–788.

Schimazek, J., Knatz, H., 1970. The influence of rock structures on the cutting speed and pick wear of heading machines (in German.). *Gluckauf*, 106, pp. 275–278.

Shao, W., Li, X., Sun, Y., Huang, H., 2014. Linear rock cutting with SMART*CUT picks. *Applied Mechanics an Materials*, 477–478, pp. 1378–1384.

Yagız, S., 2008. Utilyzing rock mass propertiesfor predicting TBM performance in hard rock condition. *Tunnelling and Underground Space Tehnology*, 24(1), pp. 64–74.

Zare, S., Bruland, A. 2013. Applications of NTNU/SINTEF drillability indices in hard rock tunneling, *Rock Mechanics and Rock Engineering*, 46, pp. 179–187.

Zhao, X.B., Yao, X.H., Gong, Q.M., Ma, H.S., Li, X.Z., 2015. Comparison study on rock crack pattern under a single normal and inclined disc cutter by linear cutting experiments. *Tunnelling and Underground Space Technology*, 50, 479–489. https://doi.org/ 10.1016/ j.tust.2015.09.002

3 Contractual Practice

3.1 INTRODUCTION

The success of a project is mainly related to whether it is finished at a scheduled time with a predetermined budget, whether it is in good quality and in international standards, or whether it is realized without any dispute between the owner and the contractor. Every country has its own realities and the type of contract selected will affect the success in the realization of the project. So, it is strictly important to understand well the advantages and disadvantages of each type of contract and the problems caused in the past between the owners and the contractors. The management of the project will depend on the type of contract to be selected. It should always be remembered that in the past there were several projects finished with unsuccessful attempts. Bearing in mind all these facts, this chapter is prepared to give a primary idea about the topic to the readers of this book.

3.2 THE MOST COMMON TYPES OF CONTRACTS BETWEEN OWNER AND CONTRACTORS

The most common types of contracts in the construction industry are summarized below. However, alternative contracting and delivery methods are also discussed by Reilly (2011), Pré et al. (2013), and Molenaar et al. (2014).

3.2.1 DESIGN–BID–BUILD

Design–bid–build (DBB) is a traditional project delivery method in which the owner contracts with separate individuals for the design and construction of a project. This delivery model has been the standard choice for infrastructure projects for a long time. It is based on separate two-party contracts: one for design – between the owner and consultant and one for construction – between the owner and contractor. This is the most cost-effective bid in a competitive market. The owner usually awards the work to the lowest bidder for a fixed price. A disadvantage of this contract form is that the design of the project has to be general enough in the tender stage, in order to attract a significant number of tenderers. Additionally, detailed design and construction occur in different phases of the project, thus generally leading to a prolonged duration of

DOI: 10.1201/9781003358978-3

the entire project. Regarding risk allocation, the owner (and his consultant) carries the full responsibility for the design of the project and for the coordination between the various parties involved in the contract. Consequently, the contractor bears a relatively small risk. In this contract, there are three main phases as given below (Huber et al. 2013; Flatley et al. 2013).

3.2.1.1 The Design Phase/Project Planning

In this phase, the owner hires a consulting engineer to design and produce bid documents, including construction drawings and technical specifications, on which various general contractors will in turn bid to construct the project.

3.2.1.2 The Bidding/Tender Phase

Bidding can be "open", in which any qualified bidder may participate, or "select", in which a limited number of preselected contractors are invited to bid. Once bids are received, the consulting engineer reviews the bids, investigates contractor qualifications, ensures all documentation is in order, and advises the owner as to the ranking of the bids. If the bids fall in a range acceptable to the owner, the owner and the engineer discuss the suitability of various bidders and their proposals. The owner is not obligated to accept the lowest bid, and it is customary for other factors including past performance and quality of other work to influence the selection process. However, the project is typically awarded to the general contractor with the lowest bid.

3.2.1.3 The Construction Phase

During the construction phase, the engineer also acts as the owner's agent to review the progress of the work as it relates to pay requests from the contractor, issue site instructions, change orders or other documentation necessary to facilitate the construction process, and certify that the project is built to the approved construction drawings.

3.2.2 Design–Build or Partial Design–Construct

The design–build (DB) approach has been used more frequently in the last few decades for infrastructure projects. In this contract model, the contractor, designers, subcontractors, and owner work together as one team to build a project that meets the expectations of the owner. The owner signs a single contract that covers all aspects of design and construction. However, the electromechanical equipment and the finishing of the structure are governed by a separate contract. This contract form is based on a lump sum award, which leads to a reduction of the total costs of the project and possible disputes between the owner and contractor. Furthermore, one contractor carries out both the project design and construction, thus reducing the total duration of the project. The main drawbacks of this type of contract are that the owner cannot influence the quality of the work and changes during the construction. The contractor is responsible for the planning and construction of the project. The advantages may be cited as switching from the standard DBB to the DB model is accelerating the whole project development process. This approach provides more opportunities to use the experience and the input of the contractor and their subcontractors to work out innovative, economical, and constructible solutions and secure cost savings in the process of design development and construction for the benefit of the project and the owner (Huber et al. 2013)

3.2.3 DESIGN–BUILD AND TURNKEY

The owner merely needs to coordinate the collaboration with third parties. Since all tunnel construction and finishing works are regulated by a single contract, the contractor generally needs to create a joint venture in order to cover all specialist fields required in the execution of the project. Due to the fact that the client awards the entire project to one single contractor, he minimizes the number of interfaces he is responsible for. However, considering that the contractor may fail to coordinate the designers, constructors, fabricators, and installers in such a diverse scope, the owner still bears the risk that he may need to take over the coordination at a later stage. Therefore, the process of selecting a qualified contractor is crucial (Huber et al. 2013).

For turnkey projects, a proposal is prepared according to the owner's specifications, and includes complete construction and delivery, releasing the owner of the interactions during the process. One can consider turnkey construction as a one-stop solution, where the contractor handles everything from communication, planning to execution with different teams to finish the project on time and within budget. This construction model is preferred by most construction companies as it gives them complete control of a project. Turnkey construction saves valuable time and money for the owner. Once the final design is approved, the construction company is solely responsible for the entire project. It is in their best interests to finish the construction well within schedule to extract an optimal return on investment or in economic terms. It is one way of relating profits to capital invested.

The disadvantages of a turnkey contract may be cited as that the entire responsibility of the contractor can also prove to be costly for the owner. The advantages of the owner can turn into a big disadvantage. Since the owner has no control over turnkey construction, there can be no changes or improvements that can be made to the project after locking on the final design. Turnkey construction allows certain contractors to get away with supplying cheap materials to maximize their profits. It is important to do a thorough background check on the construction company before getting into business with them. In certain situations, projects are stalled on account of them going over budget.

3.2.4 BUILD–OPERATE–TRANSFER (BOT)

Typical countries using BOT contracts are Turkey, Saudi Arabia, Israel, Iran. The owner benefits from two advantages of work. For one, the specified operation period motivates the contractor to design and build a fully functional object which requires a minimum of maintenance work in order to reduce operational expenses. Additionally, the owner's operating personnel are instructed by the contractor and the operation of the structure is tested for several years, allowing early detection and correction of faults. A disadvantage is that the operational management may not correspond to the wishes of the owner. In this case, the client may need to enforce a change of operational management, thus leading to disputes with the contractor and to additional expenses. The client is burdened with the risk of the consequences resulting from a poor operation of the structure caused by the contractor without being directly responsible for it (Huber et al. 2013).

An example of a very successful BOT contract is the Eurasia Tunnel Project involving the design, construction, and operation of a 5.4 km twin-deck auto tunnel beneath the Istanbul Strait including the roadways and the tunnel, with a route covering a total distance of 14.6 km. The tunnel was driven with an EPB-TBM of 15.32 diameter. ATAS, jointly owned by Yapi Merkezi and SK Engineering & Construction of Korea, obtained the authorization to build and operate for 30 years a tunnel for light motor vehicles to connect Europe and Asia. Like the other contracts, BOT also has several advantages and disadvantages (Gaille 2018). The advantages of build–own–operate–transfer (BOOT) may be cited as follows: it keeps public sector funds where they are most needed, minimizes the public cost of infrastructure development and reduces public debt, and allows for innovation. If the project was implemented by the public sector only, using innovation would not always be possible because of the costs involved. It provides a chance to bring in expertise. If the necessary expertise is not available locally, then national or international private enterprises might be brought in to create the required infrastructure. Some of the disadvantages of BOOT may be as follows: it can have higher transaction costs, it only works for large projects, and it requires fundraising to be successful. If no funds are raised to complete the project, then it won't get done. It requires strong corporate governance. One of the most common reasons for the BOOT contract failure is a lack of communication between the private and public sectors involved. When the program is being managed poorly on the private side, the public side must be able to step in and change things for the good of everyone involved.

3.3   LUMP SUM CONTRACT AND UNIT PRICE CONTRACT

A lump sum contract is a fixed-price contract. It is advantageous if the labor and materials can be estimated accurately. A lump sum amount for the entire project runs the risk that there can be unforeseen problems and developments. It has the theoretical advantage of certainty for contractor and client. However, for this form of payment to work, the contractor must bid for the job carefully. If there are cost overruns, absent agreed written change orders during construction, the contractor is at risk. Much also depends upon the expected time of the completion of the project. The longer a project continues, in a fixed bid situation, the contractor assumes the risk of increased costs of labor and materials. There are a number of advantages to employing a lump sum agreement. It is simple and cash flow is also easier to manage. There are some disadvantages to a lump sum contract as well. Lump sum contract documents can still be the source of significant disputes and disconnects. A few drawbacks include that the contractor bears the risk. Contractors bear most of the risk with lump sum contracts. If something unexpected occurs that drives the cost of the project over the lump sum fee, the contractor will bear responsibility for the excess. Since contractors do enjoy some of the project autonomy, there is less transparency for the project owner regarding how the money is spent (Zaccaria and Ferro 2019).

In a unit price contract, the total contract price is based upon the price of all the units of the work. Under a unit price contract, the contractor provides the owner with a specific price for one or more tasks or a part of the overall work that's required on the project. The owner then agrees to pay the contractor for the units that the contractor expends to complete the project. Rather than taking a look at the project as a whole and setting a price based on that finished product, a unit price contract will determine the price based on the "units" that will be required to make up that job. The costs that are commonly factored into unit prices are labor costs, material costs, overhead costs, profit, taxes, permit, and inspection costs. The unit price contract is a popular form of contract with maximum use in the field of public construction works. The advantages of a unit price contract may be cited as simple and transparent billing and flexibility. A unit price contract offers immense flexibility during the course of the project. Since these projects may last over several years, the additional need for work may come up at a later stage after the initial planning. It allows adding or expanding the project midway, which can be very helpful to the project planning and implementation team. The disadvantages of a unit price contract may be cited as accuracy in project cost and deadlines.

3.4　THE EMERALD BOOK

FIDIC (International Federation of Consulting Engineers) and ITA-AITES (The International Tunnelling and Underground Space Association) worked several years together to publish a new form of contract for tunneling and underground works, the Emerald Book. Underground construction is very much dependent on the geological, hydrogeological, and geotechnical properties of the ground, which have a defining influence on the methods required for the successful implementation of the works. In addition, the difficulty in predicting ground behavior and foreseeable conditions implies an inherent uncertainty in underground construction which gives rise to unique contractual risks regarding construction practicability, time, and cost. These risks are addressed in the Conditions of Contract for Underground Works (the Emerald Book), which was launched at the World Tunnel Congress 2019 in Naples. Several issues that the new standard form of contract should address to promote equitable risk allocation and the effective dealing with conditions typically unforeseeable in such projects. It is emphasized in the symposium that the Emerald Book will bridge an important gap. Despite all the technological developments of equipment and techniques in the field, many underground construction projects end up unsuccessful because of contractual disputes. The book brings together the result of the expertise and respectability of ITA-AITES and FIDIC in the fields of contracts and specificities for "underground constructions" https://fidic.org/node/23574. In this symposium, measuring the excavation and lining in the Emerald Book was discussed by Neuenschwander and Marulanda (2019), the claims, dispute avoidance, and dispute resolution procedure in the new FIDIC Emerald Book was discussed by Nairac (2019), and the role of the engineer in the Emerald Book was discussed by Maclure (2019).

3.5 CONCLUDING REMARKS

This contract form is based on a lump sum award, which leads to a reduction of the total. The type of contract used is one of the key factors in the success of long-running projects such as infrastructure projects. DBB is a traditional project delivery method in which the owner contracts with separate individuals for the design and construction of a project. In this contract, there are three main phases: the design phase/project planning, the bidding/tender phase, and the construction phase. The owner is not obligated to accept the lowest bid, and it is customary for other factors including past performance and quality of other work to influence the selection process. In the contract, DB or partial design–construct, the owner signs a single contract that covers all aspects of design and construction. The electromechanical equipment and the finishing of the structure are governed by a separate contract reducing the total duration of the project. This contract form is based on a lump sum award, which leads to a reduction of the total costs of the project and possible disputes between the owner and contractor. In DB and turnkey contract type, all tunnel construction and finishing works are regulated by a single contract. The contractor generally needs to create a joint venture in order to cover all specialist fields required in the execution of the project. Turnkey construction saves valuable time and money for the owner. In a BOT or BOOT contract, a private company receives authorization usually from the public sector to design, finance, construct, own, and operate, typically infrastructure projects mentioned in the contract for a period of 20 to 30 years, hoping to earn a profit. After that period, the project is returned to the public sector that was originally granted the authorization. A typical successfully terminated project in this contract type is the Avrasya Tunnel in Istanbul.

A lump sum contract is a fixed-price contract. It is advantageous if the labor and materials can be estimated accurately. However, the entire project runs the risk that there can be unforeseen problems and developments. In a unit price contract, the total contract price is based upon the price of all the units of the work. Under a unit price contract, the contractor provides the owner with a specific price for one or more tasks or a part of the overall work that is required on the project. A unit price contract offers immense flexibility during the course of the project. Since these projects may last over several years, the additional need for work may come up at a later stage after the initial planning.

REFERENCES

Flatley, A.P., Fortuna, G., Stack G., Fogaras, I.S., 2013. Evaluation of new trends in contracting and delivering underground infrastructure projects, in *Underground – The Way to the Future!*, G. Anagnostou & H. Ehrbar (eds), Taylor & Francis Group, London.

Gaille, B., 2018. Build own operate transfer advantages and disadvantages https://brandongaille.com/13-build-own-operate-transfer-advantages-and-disadvantages/, uploaded on 28th August 2022. www.netsuite.com/portal/resource/articles/accounting/lump-sum-contracts.shtml

Huber, T., Schuerch, R., Bachofner, C., Henke, F., Leu, J., Zimmermann, A., Neuenschwander, M., 2013. Comparison SIA – ITA on contractual practices, in *World Tunnel Congress 2013 Geneva, Underground – The Way to the Future!*, G. Anagnostou & H. Ehrbar (eds), Taylor & Francis Group, London.

Maclure, J., 2019. The role of the engineer in the Emerald book, in *Tunnels and Underground Cities: Engineering and Innovation Meet Archaeology, Architecture and Art*, D. Peila, G. Viggiani, & T. Celestino (eds), Taylor & Francis Group, London.

Marulanda, A., Neuenschwander, M., 2019. Contractual time for completion adjustment in the FIDIC Emerald Book, in *Tunnels and Underground Cities: Engineering and Innovation Meet Archaeology, Architecture and Art*, D. Peila, G. Viggiani & T. Celestino (eds), Taylor & Francis Group, London.

Molenaar, K., Harper, C., Yugar-Arias, I., 2014. *Guidebook for Selecting Alternative Contracting Methods for Roadway Projects: Project Delivery Methods, Procurement Procedures, and Payment Provisions*, University of Colorado Boulder, Colorado, p. 404.

Nairac, C., 2019. The claims, dispute avoidance and dispute resolution procedure in the new FIDIC Emerald Book., in *Tunnels and Underground Cities: Engineering and Innovation Meet Archaeology, Architecture and Art*, D. Peila, G. Viggiani & T. Celestino (eds), Taylor & Francis Group, London.

Neuenschwander, M., Marulanda, A., 2019. Measuring the excavation and lining in the Emerald Book, in *Tunnels and Underground Cities: Engineering and Innovation Meet Archaeology, Architecture and Art*, D. Peila, G. Viggiani & T. Celestino (eds) Taylor & Francis Group, London.

Pré, M., Thibault, J.F., Bourget, A.P.F., Russo., M, Hamaide., G, Roignot.,M, Munier, R, 2013. Presentation of the activity of the AFTES' Working Group 25, "Cost control and contractual practice", in *World Tunnel Congress 2013 Geneva Underground – The Way to the Future!*, G. Anagnostou & H. Ehrbar (eds), Taylor & Francis Group, London.

Reilly, J., 2011. Alternative contracting and delivery methods, *tunneltalk*, September 2011.

Zaccaria, L., Ferro, R., 2019. Analysis of contractual approaches in Italian railway tunnels, in *Tunnels and Underground Cities: Engineering and Innovation Meet Archaeology, Architecture and Art*, D. Peila, G. Viggiani & T. Celestino (eds), Taylor & Francis Group, London.

4 Site Set Up

4.1 INTRODUCTION

Starting the mechanized tunneling project or launching TBM is crucial in the overall process. It can be either in the mountains, like Nurdağı as described by Detlef and Bilgin (2018), Belpınar as outlined by Sakallı et al. (2019), Suruç as reported by Ilci et al. (2013) (projects in Turkey) or in urban areas where the space is limited, like Bakırköy–Kirazlı metro project (Yazıcı et al. 2017), or like several metro projects in Istanbul. There is a direct relationship between the variation of working risk level with the width of the portal or launching box, which is also a function of TBM diameter in mountains/open fields or in urban areas with limited space as shown in Figure 4.1. The contractors usually prefer to set up the portals of mechanized tunneling projects in the wide space area crossing the mountains or open fields where they also construct offices, warehouses, machine shops, dormitories of laborers, and stockyards for segments. In this case, the working risk is due to moving heavy machines, like cranes, lorries, forklifts, and the laborers charged with the assembly of the TBM, gentry, band conveyor, and ventilation fans or power plants.

In densely populated urban areas, the job sites are located close to buildings, monuments, and urban structures. All job sites are contained within station sites, being long in size. As it happens, almost none of the surface area on the job site ends up being available for TBM delivery, assembly, and launching. When the TBMs are being delivered and lowered down the shaft, station and other construction activities are occurring simultaneously and consume nearly all the available area. Since there is no room for the backup gantries at the bottom of the shaft, they are temporarily arranged on the surface in a side-by-side fashion. Even this small 10 m × 40 m area is coveted and quickly occupied as soon as the tunnel has advanced far enough for the gantries to be lowered. At one job site, in particular, an entire section of the Singapore River had to be moved so that a station site and tunnel could be built directly under it Smading (2014).

If the job sites are located in an area characterized by several historical buildings like Rome, technical and operative choices are imposed, to reduce the job site dimensions. Once in Singapore, all of the machines were assembled and launched from the bottom of the shafts. Since there is no room for assembling the entire backup at the bottom of the shaft, the backup is assembled on the surface and all the services

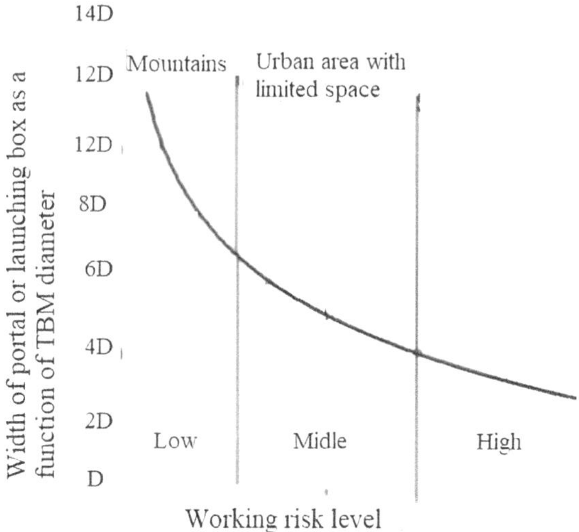

FIGURE 4.1 The variation of the working operational risk level with the width of portal or launching box with TBM diameter in mountains/open fields or in urban areas with limited space.

(electric supply and control, hydraulic, water, grout, foam, bentonite, polymer, etc.) are connected to the machine by a 130-m long umbilical cable/hose connection. Once the machine has bored about 80 m, it is stopped and the backup is disassembled, lowered into the shaft, and reassembled behind the TBM Smading (2014).

The critical aspects of setting up the portal or launching shaft are well emphasized in setting up the portal and launching shaft in TBM tunneling. There should be enough room for the assembly and launching of the TBM. When crossing the mountain land reclamation and in the case in urban areas, a shaft of several to 10 m in depth might be necessary. Special measures should be taken during the planning stage of portals and shafts. These are as follows: exposure to flooding, fire, and theft during the period of storage, exposure to damage caused during the assembly due to the lifting and movement of the heaviest components, exposure to flooding or landslides during the assembly of the machines. There should be enough room for the assembly and launching of the TBM. The bottom shafts should, in this case, also be equipped with de-watering pumps. In order to shorten overall delivery times to supply these machines, some suppliers have recently introduced the so-called Onsite First Time Assembly (OFTA), consisting of operating only one assembly directly at the site. These aspects of tunneling with TBM are well described in IMIA Working Group Paper WGP 60 (09) (2009) Tunnel Boring Machines. The following paragraph is a summary about this respect.

As Romani et al. (2019) emphasized, in an urban environment, the job sites require the reduction of their sizes as much as possible, with significant consequences on the design and operational choices, in particular those related to the TBM excavation of the tunnels. In their case, starting from that shaft, a gallery was constructed for a length of about 385 m. They reported that the Via Sannio shaft is 62 m long and 22 m large: it is totally undersized as compared to those generally needed to enable the operation of these TBMs, which are equipped with a 100 m backup. The two TBMs (of about 450 tons each) were entirely assembled on the surface and lowered through a temporary slot down to the bottom of the shaft by a crane of 750 t. Once laid on purpose-built steel structures, and placed at the bottom of the shaft, the TBMs were moved as far as the head diaphragm wall to start the excavation. Two false tunnels for the launch of the TBMs were constructed in the shaft. After the passage of the TBM backup, the false tunnel structure is demolished. The dimensions of the shaft are also required to start the excavation with only four out of seven muck cars installed at the rear of the TBM.

As Tirolo et al. (2013) described, a launch box may have dimensions of 244 m long by 18.9 m wide and a depth of 19.8 m. As Zhan and Xuejin (2014) emphasized, the launching of a TBM is very important in the tunnel construction procedure, but it also has great risk. Water and slurry may burst into the shield in the case of an inappropriate launching process, resulting in the collapse of the launching shaft.

Taken from the statements given earlier, this chapter will summarize the basic principles to dimension the portals in mountains/open fields and launching boxes in urban areas with some examples from the projects in Turkey.

4.2 DESIGNING THE PORTALS IN MOUNTAINS/OPEN FIELDS

Suruç Tunnel, with a length of 17,185 m, is the longest tunnel ever excavated in Turkey. It is planned to irrigate the Eastern Anatolian part of Turkey. The expected water flow rate is planned to be 90 m³/s and irrigates an area of 94,814 ha. The excavation of the 7.88 m diameter tunnel started with double shield hard rock TBM in September 2010 and finished in 2014. The main rock formation in the area is the Gaziantep Formation of Eocene-Oligocene aged karstic chalky clayey limestone. Clay bands sometimes have 50–75 cm thickness. Groundwater was encountered in the majority of drill holes. A double shield hard rock TBM provided by Selli was used in Suruç Tunnel. The main characteristics of TBM are as follows: TBM diameter, 7.88 m; cutting head power, 4,500 kW; maximum torque, 4,000 kNm; maximum thrust, 20,000 kN; weight, 1,080 t (Ilci et al. 2013). The portal is seen in Figure 4.2 with dimensions of 100+ lengths of the gentry. The portal is designed for assembling TBM, warehouses, workshops, offices, segment stockyards, and dormitories. There was a dispute between the job owner (DSI, State Water Organization) and the contractor Ilci about the over-excavation of the portal concerning the payment of dues. A model was created to justify the dimensions of the portal mainly based on safety concerns by Bilgin (2013). It is proved that the results obtained from the model created were in good agreement with the current dimensions of the portal. On 22 October 2012, flooding occurred in the portal with a water ingress of 28 m³/s, 120,000 m³ of water accumulated in the portal with 4 m water height, and 80,000 m³ of water entering into the tunnel. Thirty-six workers were inside the tunnel. It is apparent that if the portal was not designed in such tolerated dimensions, it would be a disaster for the workers. The flooding of portal in Suruç Tunnel on 22 October 2012 is seen in Figure 4.3.

FIGURE 4.2 The portal of Suruç Tunnel, Turkey. (From the archive of the author, Bilgin 2013.)

FIGURE 4.3 The flooding of portal in Suruç Tunnel on 22 October 2012. (From the archive of the author, Bilgin 2013.)

FIGURE 4.4   AMR II inlet portal before flood. (Langmaack et al. 2010.)

Another example of the flooding portal is the Andhra Pradesh/India irrigation tunnel project. Three 10-m diameter, double shield TBMs were used for the excavation of both Alimineti Madhava Reddy (AMR) and Veligonda tunnels – part of a large irrigation scheme by the Andhra Pradesh government. A major flooding occurred on 2 October 2009, at the AMR Inlet, when the job site was hit by a 100-year monsoon which flooded the job site and covered the TBM and backup. The equipment was under water for approximately ten days until the water was pumped out. Later, the equipment was completely refurbished by TBM personnel at the job site (Harding 2010), AMR II Inlet portal before and after the flood, as is seen in Figures 4.4, 4.5, and 4.6 (Landmark et al. 2010; Wallis 2009).

The model created to dimension a portal to be opened in mountains/in open fields is given below (Bilgin 2013):

$$W_{Portal} = D_{TBM} + 2\left(k_1 + k_2 + k_3\right) + 2(W_c)^n + 2(W_L)^n + W_f + W_{etc} \tag{4.1}$$

In this equation:

W_{PORTAL} is the width of portal

D_{TBM} is the diameter of TBM

FIGURE 4.5   AMR II inlet portal after flood. (Langmaack et al. 2010.)

k1 is the distance left for safety reasons at the entrance of the tunnel, in our case k1 = 4 m

k2 is the distance left for safety reasons between crane and TBM and crane lorries, in our case k2 = 4 m

k3 is the distance left between TBM and crane, in our case k3 = 5 m

$(W_C)^n$ is the width of crane with operation distance, in our case $(Wc)^n$ = 15 m

$(W_L)^n$ is the width of heavy lorries with operation distance, in our case $(W_L)^n$ in our case $(W_L)^n$ = 10 m

$n = f(D_{TBM})$, n changing from 0.5 for TBMs having diameter under 4 m and 1 for TBMs having diameter greater than 7 m (4.2)

Wf is the width of forklifts and similar equipment with operation distance, in our case Wf = 3 m

W_{etc} = Width spared for (warehouse + workshop + offices + segment stock yard + dormitories), in our case 11 m

Using Equations (4.1) and (4.2), W_{PORTAL} is calculated as 7.88 + 2(4 + 4 + 5) + 2 × 15 + 2 × 10 + 5 + 10 = 99 m

A second example of designing a portal in mountains/in an open field, the Belpınar Tunnel

FIGURE 4.6   Recovery of TBM from flooded portal. (Harding 2010.)

Belpınar irrigation Tunnel of 5,450 km in length was opened with an EPB-TBM having a diameter of 6.82 m. The tunnel has a complex geology with serpentinites and peridotites, highly fractured in most cases with water ingress reaching up to 450 L/s. The excavation of the 5,450-m long Belpınar Tunnel of 6.82 m in diameter, which connects Saglık Plain to Emek Plain, was started on 24 December 2014 by an EPB-TBM and completed on 6 November 2016, which was within the scope of construction of 48,564-m long Kilavuzlu Irrigation Main Channel. Irrigation water will be provided to a total area of 55,536 ha in the city of Gaziantep and an additional area of 33,400 ha in Amik Plain. The portal in Belpınar Tunnel/Turkey is seen in Figure 4.7 (Sakallı et al. 2019).

4.3   SITE SET UP WITH TRENCH ENTRY FOR TBMS

Launching boxes is inevitable in urban areas for TBMs due to limited space. The launching box designed with tick wall tubes as struts for Bakırköy–Başakşehir–Kirazlı

FIGURE 4.7   Portal in Belpınar Tunnel/Turkey. (From the archive of the author, Bilgin.)

twin metro tunnels is seen in Figure 4.8. Struts and bored piles are necessary to be used on soft ground as found in urban areas in Istanbul to overcome lateral deformation. The owner of the project is the Istanbul Metropolitan Municipality and the contractor is Aga Enerji. The construction of the works concerns twin tunnels of 2 × 6.2 km in length with two TBMs having a diameter of 6.57 m. Station tunnels are excavated with conventional tunneling methods. The dimensions of the launching box are 28 × 185 m with a depth of 23.5 m. The O-ring sealing system was used for TBM break by Yazıcı et al. (2017). The Bakırköy Formation found in the area consists of weathered clayey limestone with karstic cavities and permeable structure. The launch box is almost 100 m far away from the Marmara Sea, and water ingress due to the high underground water level was a great concern when designing the launching box.

Another launching box designed with a steel structure for a single-line metro project in North Melbourne is seen in Figure 4.9.

4.4 SHAFTS ENTRY AND PORTALS IN URBAN AREAS

New Istanbul Airport is constructed over an area of 76.5 million square meters on a former open-cast coalfield located on the European side of Istanbul, 35 km northwest of the city center toward the Black Sea coast. The airport hosted 23.4 million passengers in 2020 despite the serious decrease in the number of passengers at all airports worldwide due to the Covid-19 pandemic. It is planned to offer flights to more than 350 destinations globally. The airport is currently connected to the city by a fleet of buses, but the metro will make its way to the new airport by 29 October 2021, the date of the

FIGURE 4.8 Launching box for Bakırköy–Başakşehir–Kirazlı twin metro tunnels. (From the archive of the author, Bilgin.)

FIGURE 4.9 The site of the future Arden Station, 131 Laurens Street North Melbourne.

Source. https://metrotunnel.vic.gov.au/about-the-project/news/2019/tunnel-boring-mach
ine-tour

FIGURE 4.10 The shafts at Gayrettepe Station on the Gayrettepe line in Istanbul's new airport. Photograph by Ümit Kılıç. (Bilgin and Acun 2021.)

foundation of the Turkish Republic. There are two lines with two tubes each: Line 1 with a length of 37.5 km from Gayrettepe to the airport, and Line 2 with a length of 35 km from Halkali to the airport. The construction of the second line from Halkali to the airport is underway. The first line, from Gayrettepe to the airport, has nine stations and a total (twin tube) length of 75 km (of which 69 km was bored by TBM). In order to excavate the long TBM tunnel drives, ten EPB-TBMs were used (6.56 m diameter) from four different TBM suppliers. The geology along the TBM tunnel routes is complex, with several transition zones, shear zones, and faults. The line is planned for 150,000 passengers per hour, with travel times between the city centre and the airport taking around 24 minutes. One of the seven shafts designed for the project is seen in Figure 4.10. Approximate dimensions/areas of job sites and related areas are given in Table 4.1. Figure 4.10 is a general view of the shafts at Gayrettepe Station on Line 2.

TABLE 4.1

Approximate Dimensions/Areas of Job Sites and Related Areas

Job Site and Total Area (m²)	Launching Box/Shaft w × L × h (m), (m²)	Dormitory (m²)	Chiller (m²)	Muck Area (m²)	Grout Central (m²)	Ware House (m²)	Segment Stock (m²)
Kağıthane 15,000 m²	23.80 × 80 × 23	–	250	950	450	400	500
Hasdal 44,000 m²	29 × 200 × 23	5,000	200	1,200	450	400	700
Kemerburgaz 39,000 m²	23.80 × 200 × 23	4,000	250	1,050	450	400	900
İhsaniye 42,000 m²	23.80 × 200 × 23	5,000	200	1,300	450	400	500
Yenimahalle 15,000 m²	24 × 180 × 29	4,00	150	1,570	450	400	400

Source: From the archive of the author, S. Acun.

Note: Yenimahalle job site is from another Metro Project in Istanbul, Mahmutbey–Mecidiyeköy.

FIGURE 4.11   Kağıthane job site/launch box. Photograph by Ümit Kılıç.

This figure shows clearly how much a shaft of a metro project in a densely populated city may be difficult to construct having a high working risk level. Kağıthane, Hasdal, Kemerburgaz, and İhsaniye job sites/launching boxes, which are the subject of this section, are from Line 1 of Gayrettepe–Istanbul Airport Metro Project.

Figures 4.11, 4.12, 4.13, and 4.14 are areal views of four job sites/launching boxes and related areas with some brief explanations. We believe that this information may be a guide for the readers to visualize the management of sites/launching boxes in a big metro project.

4.4.1   Kağıthane Job Site/Launch Box

Two TBMs (Herrenknect) of 6.7 m diameter worked toward Hayrttepe/city centre. Two hundred tunnel staff worked on this job site. The local school and the river, which exist in this area, limited the space to 15,000 m².

4.4.2   Hasdal Job Site/Launch Box

Two TBMs (Lovsuns) excavated toward Kağıthane station and 200 people worked in this area. A space of 5 × 4 m in dimension was set at the bottom of the shaft, and a space with the same dimension was set above the shaft for muck drainage and a treatment/refining area was also set on the job site. Systematic chemical analysis was carried out on the samples taken from the muck, and careful planning of an environmental protection program was applied in this area.

FIGURE 4.12   Hasdal job site/launch box. Photograph by Ümit Kılıç.

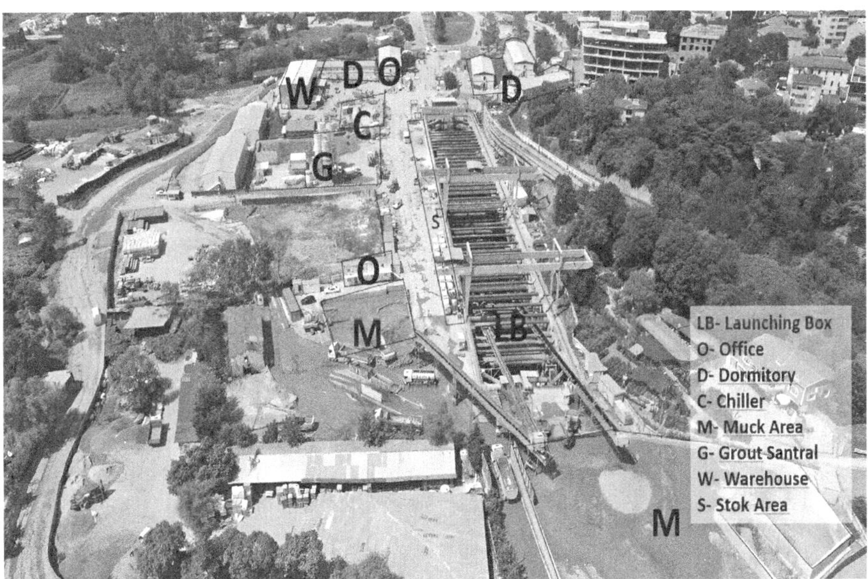

FIGURE 4.13   Kemerburgaz job site/launch box. Photograph by Ümit Kılıç.

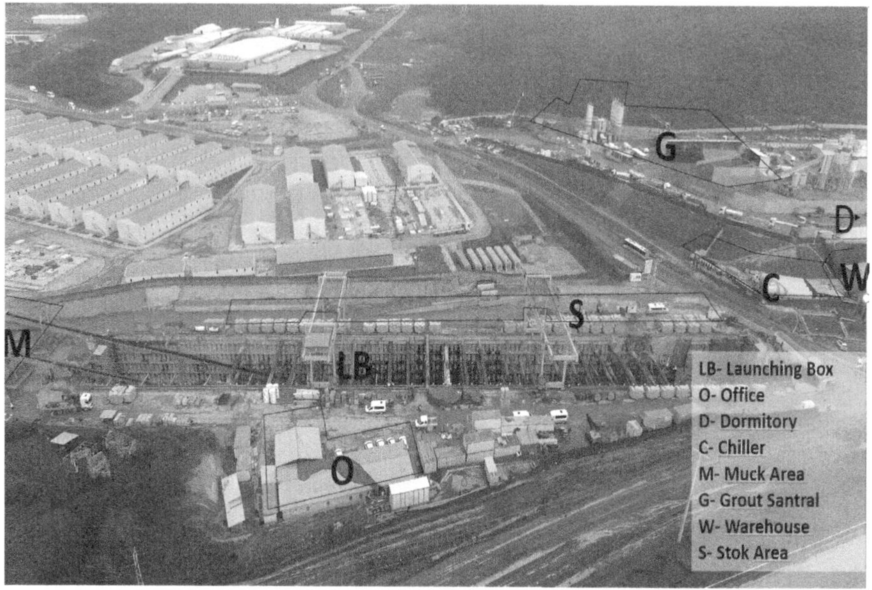

FIGURE 4.14   İhsaniye job site/launch box. Photograph by Ümit Kılıç.

4.4.3   KEMERBURGAZ JOB SITE/LAUNCH BOX

In this work site, 330 staff were employed. Two NHI TBMs worked on the right side and one Terrateck TBM and one Lovsuns TBM worked in the opposite direction. Launching boxes, offices, dormitories, chiller, grout units, warehouse, and stock area were located in this area.

After an advance of 2,500 m toward the airport, chiller, ventilation units offices, grout units, and segment stock area were moved to Göktürk station in order to shorten the distance travelled with multi-service vehicle.

4.4.4   İHSANIYE JOB SITE/LAUNCH BOX

A group of four Herrenknecht TBMs, consisting of two TBMs moving in the left direction and the other two TBMs moving in the opposite direction, worked in this area and a total of 360 staff (engineers, technicians, and workers) were employed in this area. Twenty-four hours of working time were scheduled in all areas.

4.5   OFTA (ONSITE FIRST TIME ASSEMBLY)

TBMs are usually assembled in factories, where the components are assembled and tested, then disassembled and shipped to the job site. Delivery of a machine can often be the critical path affecting project schedule, cost, manpower, etc. However, in order to shorten overall delivery times to supply TBMs, OFTA has been developed and used efficiently on several projects around the world by some companies. Home (2010) has also cited early indications of OFTA benefits as described in Table 4.2.

TABLE 4.2
Early Indications of OFTA Benefits

Project Machine	Machine Type/Diameter	Assembly	Savings
Niagara Tunnel	14.4 m Main Beam TBM	17 weeks	3–4 months
AMR Water Tunnel	2 × 10 m Double Shield TBMs	16 weeks	3–4 months
Mexico City Metro	10.2 m EPB-TBM	18 weeks	2–3 month

Source: Home (2010).

FIGURE 4.15 Onsite First Time Assembly in Bahçenur High-Speed Railway Tunnel Project. (Detlef and Bilgin 2018.)

One of the typical examples of OFTA from Turkey may be cited as Bakçenur High-Speed Railway Tunnel Project. The tunnel is situated in Gaziantep Province, in Southeastern Turkey, which is an important center of agriculture and trade divided into nine districts. With a population of nearly 1.7 million, the province is overhauling its public transportation with a rail line between the town of Bahçe and the Nurdağı district. The Bahce–Nurdağı Railway Tunnel consists of two parallel 10.1 km tunnel. The Single Shield TBM of 8.0 m was assembled at the job site using OFTA. TBM assembly started the job site on 15 July 2015 and terminated on 21 September 2015, including gantry assembly and testing all systems taking 68 days. The job site location about 32 km from the Syrian border complicated shipping of some parts and transporting TBM parts started in May 2015 (Detlef and Bilgin 2018). OFTA in Bahçenur High-Speed Railway Tunnel Project is illustrated in Figure 4.15. The advantages of OFTA activities and time schedule for this are clearly seen in Table 4.3. Transport of the TBM parts in this mountain and the difficult area took 122 days and the assembly of the TBM took only 68 days.

Assembly of TBM in the Gerede Water Conveyance Tunnel is unique in its operational management. Harding and Alpagut (2020) described the following operation:

The December 2018 breakthrough of a 5.5 m diameter hybrid-type single shield/ EPB TBM at the Gerede Water Transmission Tunnel in Central Turkey was a feat of

TABLE 4.3
OFTA Activities in Bahçenur High-Speed Tunnel

Activities	May	June	July	August	Sept.
Transport of Parts, 105 days		——	——————————————		
TBM OFTA Assebly, 68 days			——————	———————	
Support and main bearing, 7 days			—		
Front shield and cutterhead support, 2 days			—		
Front shield parts, 13 days			——		
Half down shield parts,5 days				—	
Operator units,4 days				-	
Down half of the tail shield, 3 days				-	
Bridge and carrier,3 days				—	
Support which is behind of the shield, 7 days				—	
Inclined circle, 2 days				-	
Erector, 7				—	
Upper shield parts, 14					——
All systems on the bridge, 7 days					—
Front shield and cutterhead support, 4 days					—
Hydraulic systems of TBM, 4 days				—	
TBM electrical parts, 31 days				————	—
Gantry assembly, 52				————	—
1-9 gentries, 14 days				—	
Reassembly of systems if necessary, 16 days				—	
Gentries to TBM, 7 days					—
Conveyors, 7 days				—	

Source: From the archive of N. Bilgin.

TABLE 4.3 (Continued)
OFTA Activities in Bahçenur High-Speed Tunnel

Ventilation systems, 9 days	—
TBM electrical and hydraulic systems, 15 days	
Data acquisition and visual systems,	———
6 days	—
Test, 9 days	———
Tests, 7 days	—
Getting reedy of TBM, 2 days	—
Segment production (Carrausel), 87 days	————————

modern construction. The 9 km leg was the final section of the 31.6 km long water supply line bored through what is widely considered to be Turkey's most challenging geology. The project was originally started with the contractor selecting three Double Shield machines, which were procured and supplied without Robbins involvement. When two of the machines became stuck and were unable to continue, the solution of the hybrid-type TBM was developed to complete the rest of the tunnel. The Robbins XRE TBM was launched in summer 2016, Crews excavated a bypass tunnel to one side of the stuck Double Shield (TBM-3), and the Robbins TBM components were walked in through the south portal. The machine was assembled using Onsite First Time Assembly (OFTA) in an underground launch chamber. The logistics of getting components through the existing tunnel were very challenging, as the assembly chamber was 7 km from the portal. The water inflow made it difficult to get the materials to the machine. To overcome this, custom flat cars equipped with hydraulic lifts were used as seen in Figure 4.16. Assembly chamber configuration is illustrated in Figures 4.17 and 4.18.

FIGURE 4.16 TBM assembly in the chamber and custom flat cars. (Harding and Alpagut 2020.)

FIGURE 4.17 TBM assembly in the chamber in Gerede Tunnel, Turkey. (With kind permission of Timur Koloğlu, from Kolin Construction Company.)

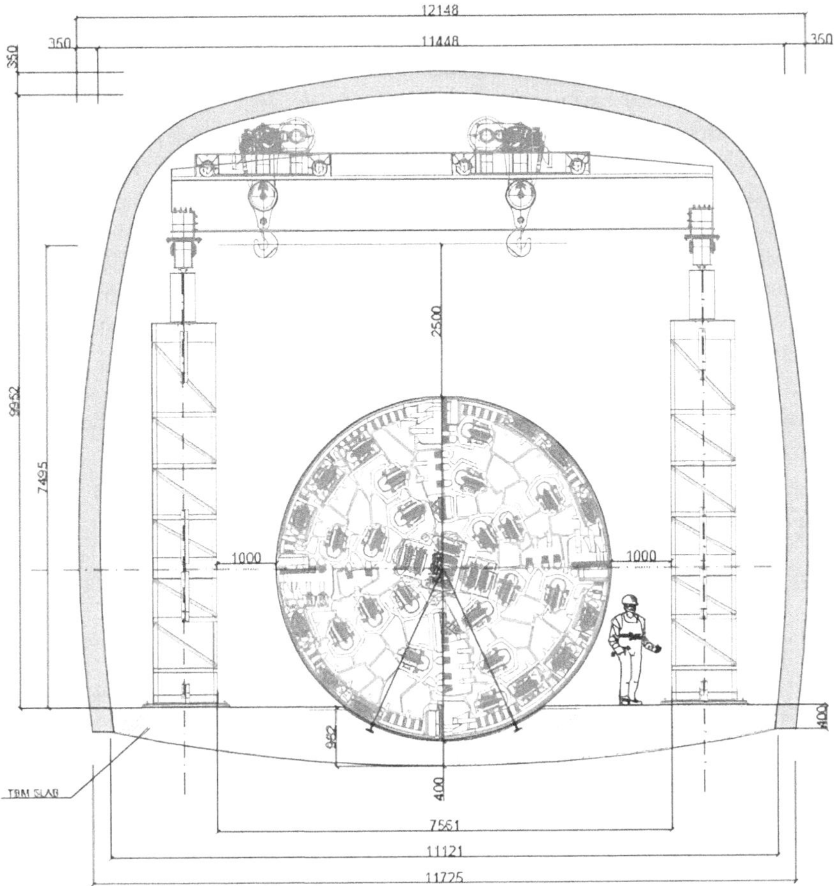

FIGURE 4.18 Assembly chamber configuration. TBM assembly in the chamber and custom flat cars. (Harding and Alpagut 2020.)

4.6 CONCLUDING REMARKS

Planning of job sites, portals, or launching boxes is of prime importance in the success of mechanical excavation. In mountains or in open fields where there is a sufficient place, usually, the launching box, offices, muck area, segment stock area, grout plant central, chiller, and warehouse are placed within the same area. This optimizes the moving of workers, equipment, lorries, cars, etc. In a metro project with twin tunnels of 7 m excavation diameter, the total area needed changes between 15,000 to 44,000 m^2 for safe and efficient operation, as in the Istanbul Airport Project. In such a project, the areas needed for launching box changes (23/29–180/200 m), offices (400 m^2), muck area (950–1,200 m^2), segment sock area (500–900 m^2), grout central (450 m^2), chiller (200–250 m^2), an warehouse (400 m^2). In this project, from the total length of 75 km tunnel, 69 km of tunnel was opened by a total number of ten TBM. However, for highly populated urban areas, it is not possible to find such wide places. For example, in Via Sannio shaft, the dimension of the launching box is 62 m long and 22 m large, and a false tunnel is excavated at the bottom of the shaft for the assembly of TBM and gantry. It should be borne in mind that the risk of operational safety increases with the decreasing size of job sites.

Another important point is that OFTA was recently introduced by some manufacturers. TBMs have been traditionally assembled in factories, where the components are assembled and tested, then disassembled and shipped to the job site. Delivery of a machine can often be the critical path affecting project schedule, cost, manpower, and other factors. As Home described, in Niagara and AMR tunnel OFTA saved 3–4 months in the project schedule. It is also proved that a TBM of 5 m in diameter may be transported and assembled within the tunnel 7 km away from the portal, as happened in the Gerede Tunnel, where the water ingress into the tunnel was at a high level.

We believe that the examples given in this chapter will serve as an example of designing the job sites, to diminish the time for transportation of TBMs and assembly at the site for future tunneling projects carried out in similar conditions.

REFERENCES

Bilgin, N, 2013. *Technical report written for Ilci, the contractor of Suruç Tunnel*, Istanbul Technical University, 11 September 2013, p. 2018.

Bilgin, N., Acun, S., 2021. Learning from Istanbul, *Tunnels and Tunnelling, International Edition*, August 2021, pp. 18–22.

Detlef, J., Bilgin, N., 2018. Turkey's hardest rock: The Bahce-Nurdag High Speed Railway *Tunnel*. 4, pp. 22–28.

Harding, D., 2010. Tunnel boring machines used for irrigation in Andhra Pradesh, India, *ITA-AITES World Tunnel Congress*, 14–20 May, Vancouver, Canada.

Harding, D., Alpagut, Y., 2020. Tunneling through 48 fault zones and high water pressures on Turkey's Gerede water transmission tunnel, *ITA-AITES World Tunnel Congress, proceedings of WTC2020* and 46th General Assembly, Kuala Lumpur, Malaysia, 11–17 September 2020, pp. 759–766.

Home, L., 2010. To build a tunnel boring machine: Why assembly on location www.therobbins company.com

Ilci, N., Temel, M., Sezgin, S., Akpınar, T., Guarasio, S., Polat, C., Bilgin, N., 2013. Clogging and squeezing effect of marl-clayey limestone on the performance of a hard rock TBM in Suruc Tunnel, Turkey, in *World Tunnel Congress 2013 Geneva Underground – The Way to the Future*, G. Anagnostou & H. Ehrbar (eds), Taylor & Francis Group, London.

IMIA Working Group Paper WGP 60 (09), 2009. Tunnel boring machines, *IMIA Conference Istanbul*, 2009, IMIA, the International Association of Engineering Insurers held its 42nd Annual Conference at Istanbul from 28 to 30 September 2009.

Langmaack, L., Grothen, B., Jakobsen, P.D., 2010. Anti-wear and anti-dust solutions for hard rock TBMs, *Tunnelling Association of Canada, WTC* Vancouver.

Romani, E., D'Angelo, M., Foti, V., Magliocchetti, A., 2019. The T3 stretch of Line C in Rome: TBM excavation, in *Tunnels and Underground Cities: Engineering and Innovation meet Archaeology, Architecture and Art*, D. Peila, G. Viggiani & T. Celestino (eds), Taylor & Francis Group, London.

Sakallı, M., Talu, D., Bilgin., Aksoy, I.H., 2019. Problems associated with an EPB-TBM in a complex geology with serpentinites and peridotites in Turkey, *World Tunnelling*

Conference, Naples, Tunnels and Underground Cities: Engineering and Innovation Meet Archaeology, Architecture and Art, pp. 4218–4224.

Smading, S., 2014. Logistics of limited space urban tunnelling at Singapore's mega metro, *Proceedings of the World Tunnel Congress 2014 – Tunnels for a Better Life*, Foz do Iguaçu, Brazil.

Tirolo, V., Maxwell, T., Parikh, A. (2013). Construction of a TBM launch box in a complex urban environment. *North American Tunneling 2012 Proceedings*, 65, pp. 50–55.

Wallis, S., 2009. Serious flood setback for TBM project in India, *Tunnel Talk*, October.

Yazıcı., H.A., Gümüş, U., Durmuş, Y.G., Okkerman, M.B., Budak, C., Gül, Ş, 2017. 0-ring sealing system for TBM break in: *International Tunneling Symposium in Turkey organized by Turkish Tunnelling Society: Challenges of Tunnelling*, pp. 123–131.

Zhan, W., Xuejin, Lv., 2014. Study on TBM launching in parallel with small spacing, *Geo-Hubei International Conference on Sustainable Civil Infrastructure*.

5 Tunnel Transport for Mechanized Tunneling

5.1 INTRODUCTION

In a tunneling operation, the muck transportation with a belt conveyor or locomotive affects the overall performance. Khetwal et al. (2021) studied two different muck transportation systems, conveyors and locomotives, under similar geological conditions to quantify the effect of the selection of muck transportation systems on machine utilization. They concluded that the conveyor belt shows an average increase of 3% in machine utilization as the length of the tunnel increases in hard rock. However, in some cases, this average may go up to 5% as Rostami (2016) pointed out. This was also pointed out by Farrokh et al. (2013), Namlı et al. (2013), and Namlı and Bilgin (2017) claiming that in a TBM drive, if the transportation distance is between 0 and 3 km, the stoppages in locomotive transport are 7% and this value decreases down to 5% in belt conveyors in overall performance. For distances greater than 3 km, the stoppages by locomotive are 10%, decreasing down to 7% in belt conveyors.

Mucking out with a conveyor belt is a common technique for full-face tunnel boring with TBM, and is being currently used in several projects all over the world. However, compared to locomotive transport it is also found efficient in drill and blast excavation methods (Fossati et al. 2017).

It is always the contractor's decision, but generally, if the tunnel is longer than 1.5 km in length, conveyors are used. If it is shorter, then muck cars are usually preferred. With small diameter TBMs, although there is an opportunity to develop smaller belt conveyors, in practice belt conveyors smaller than 60 cm wide are not used. On the other hand, some contractors because of their lower investment cost compared to belt conveying prefer locomotive transport.

With TBM drives, the belt conveyor is extended continuously while the TBM progresses without stopping. To realize this, the installation of a special frame elongation station onto the TBM's trailers is necessary, as well as the installation of belt storage in the drive unit area. Horizontal or vertical belt storage may be used; the vertical model is preferred in urban areas where little space is available.

Bringiotti et al. (2019) cited the advantages of using such belt conveyors as, the reduction of costs in long tunnels, environmental empact, safety, and considerable energy saving in comparison with gas oil consumption in rail transport. Material excavated from the TBM is continuously moved in the tunnel and/or from the

DOI: 10.1201/9781003358978-5

excavation shaft, there is no interference on the road between vehicles going out (mucked material) and in (segment transportation, injection material, and personnel). There is no auto-repair garage for assistance and maintenance of the locomotors, no tankers or re-fuelling plants, optimization and minimization of the spare part area, and fewer personnel to handle the muck as a whole (drivers, mechanics, maintainers, safety systems inspectors, etc.). They also cited the disadvantages of rail transport such as waiting for locomotives, setting up switches, derailing of locomotives, and additionally, the need to upsize the ventilation if diesel locomotives are used. With the belt conveyor system, the only time tunneling stops is to add a belt. However, contractors in some cases prefer rail transport because of its lower initial cost compared to belt conveyors.

Due to the importance of the muck transportation systems to the overall efficiency of mechanical excavation, this chapter will be devoted to muck transportation with belt conveyors, locomotives, and material transport with multiple-service vehicles (MSVs). First, some general information on the subject, later some examples from different projects and numerical examples of the selection of the transport systems will be given in order to help the readers understand the best choice in practice.

The suppliers usually make calculations for muck transportation systems, belt conveyors, or locomotive transport on a given project and make some suggestions. However, the designer must also make some primary calculations, which will allow him to suggest some recommendations for a proper muck transport system. This will allow the project team to have primary data for the management of the project, feasibility, time scheduling, and cost calculations. In light of this, the following chapter is devoted to giving some primary information on the subject and one numerical example from Turkey of belt conveyor for the Mahmutbey-Mecidiyeköy Metro project.

Regulations related to tunneling operations on belt conveying, permanent haulage track and rolling stock, and MSVs are within the definition of mechanical apparatus, which should be included in the mine manager's scheme of maintenance and safety required by those regulations. It is the responsibility of the mechanical engineer to make sure that the mechanical engineering staff supervises or affects the installation, examination, and maintenance in safe working order of all permanent tracks. Sufficient competent staff should be available to allow this to be done.

5.2 MUCK TRANSPORT WITH BELT CONVEYORS

5.2.1 HORIZONTAL BELT CONVEYORS

Figures 5.1 and 5.2 show typical arrangements used for belt conveyors. The conveyor is equipped with a tension system to allow the transmission of power between the drive pulley and the belt. There are three principal devices of belt tensioning (with screws or hydraulic) for conveyors as defined by Bringiotti et al. (2019). If the tunnel is bored using a TBM, then the belt conveyor could extend continuously while the TBM progresses, without stopping. In this case, the installation of a special frame elongation station onto the TBM's trailers is necessary, as well as the installation of belt storage in the drive unit area. In both cases, it is necessary to dimension the belt

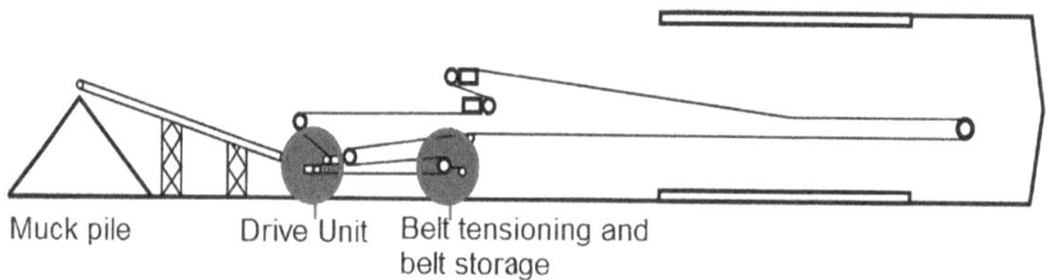

FIGURE 5.1 Typical arrangements used for belt conveyors. Courtesy of Komatsu-Joy.

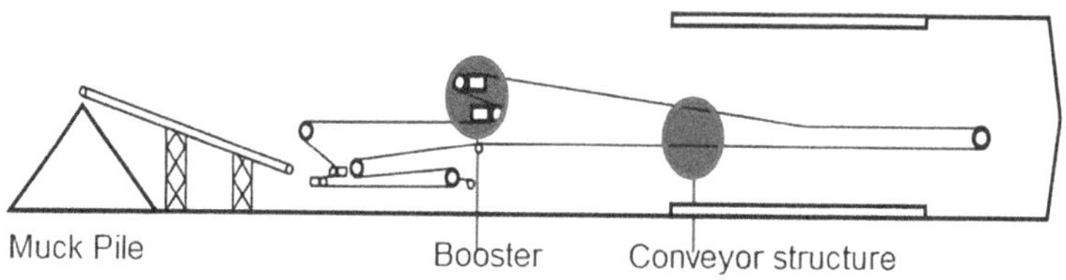

FIGURE 5.2 Typical arrangements used for belt conveyors with booster. (From the brochure released by Komatsu-Joy.)

conveyor, taking into account the specifications met along the tunnel route, such as horizontal or vertical curves, and slopes.

In long tunnels or in tunnels having curvatures, boosters are used to reduce the overall belt tension. In this way, the excavated material can be transported safely long distances and around the tightest curves. The location of the booster drives is critical since the location directly affects the maximum tension developed in the belt. Booster drives are typically installed on the loaded side of the belt (tripper drives), but may also be required on the return side of the belt in longer tunnels. The installation of a typical belt conveyor in a tunnel having a tight curvature is seen in Figure 5.3.

5.2.2 Some Examples from Turkey on Using Belt Conveyors in Tunneling Projects

Table 5.1 gives a summary of the characteristics of the belt conveyors used in Turkey. All belt conveyors are supplied by Herrencknecht H+E except the last two supplied by Robbins. Typical views of belt conveyors from Eşme high-speed tunnel and Kargı power tunnels in Turkey are seen in Figures 5.4 and 5.5

5.2.3 A Numerical Example on the Selection of Belt Conveyors

The following calculations are made for Mahmutbey-Mecidiyeköy Metro Line which consists of 18.5 km twin tunnels with a combination of TBMs and conventional tunneling methods. Approximately half of the length of the line is excavated by 3 EPB-TBMs (Bilgin et al. 2017). The length, the gradient, and the curvature in different sections of the tunnel are given in Table 5.2. The excavation diameter is 6.6 m and

FIGURE 5.3 The installation of a belt conveyor, with courtesy of Herrencknecht.

it is expected that the maximum daily advance rate will be 25 rings or $25 \times 1.4 = 35$ m or 1,200 m^3 volume of the excavated material per day or 50 m^3 per hour which makes $50 \times 2.2 = 110$ t/h, 2.2 t/m^3 being the mean density of the excavated material (Bilgin, 1977).

There are several handbooks in the market for the design and calculations for the belt conveyors such as in Aşık (1988) and CEMA (2020, 7th edition). The methodology developed by Brook (1971) will be followed below due to its simplicity.

Calculation steps followed after Brook (1971) are as follows:

a. The selection of belt width and belt speed

Belt width (W) and belt speed (V) are two parameters necessary to calculate the belt capacity. Belt width changes with the size of the material on the belt and typical values for the belt width change from 0.8 to 1.0 m and belt speed is 3.5 m/sec as seen from tables given in standard books.

Belt width and speed are selected to be 0.95 m and 3.0 m/sec consulting tables given by Aşık (1992).

b. The maximum cross section (A) of the muck pile on the belt

As a rule of thumb, this area may be calculated as $W^2/11$ for big-size lumps, Brook (1971). However, it is proved that in EPB tunneling the area may be as low as $W^2/30$

In this case $A = 0.95^2/30$ or $A = 0.041$ m^2

c. Checking the suitability of the belt capacity (Q) for the example given

TABLE 5.1
Some Examples of the Belt Conveyors Used in Turkey

Project	D (m)	L1 (m)	L2 (m)	W (cm)	V (m/sec)	Q (t/h)	P (kW)
Atakoy–Ikitelli Metro	6.6 EPB	5,500	5,367	800	3.5	450	750
İstanbul Airport Line 1	6.6 EPB	4,200	4,200	800	3.5	450	400
İstanbul Airport Line 2	6.6 EPB	3,560	3,650	800	3.5	450	400
Goztepe Metro Line 1	6.6 EPB	6,400	6,400	800	3.5	450	675
Göztepe Metro Line 2	6.6 EPB	6,400	6,400	800	3.5	450	675
Götepe Metro Line 3	6.6 EPB	6,240	6,240	800	3.5	450	475
Göztepe Metro Line 3	6.6 EPB	6,240	6,240	800	3.5	450	475
Esme Railway	13.8 XRD	3,400	3,400	1,000	3.5	1,600	710
Kemerburgaz–Hasdal	6.6 EPB	5,090	3,500	800	3.5	450	400
kemerburgaz ihsaniye	6.6 EPB	5,855	5,855	800	3.5	300	600
Istanbul Airport Ihsaniye Line 1	6.6 EPB	9,860	9,860	800	3.5	300	875
Istanbul Airport Ihsaniye Line 2	6.6 EPB	7,220	7,220	800	3.5	300	875
Istanbul Airport Hasdal	6.6 EPB	5,330	5,330	800	3.5	300	600
Kargı Power tunnel*	9.8 DS	11,700	11,900	950	3.5	1,500	370
Bahçe–Nurdağı Railway	8.0 SH	9,700	88,500	1,000	3.5	650	600

Source: From the archive of the author, N. Bilgin.

Note: L1: tunnel length; L2: length of the belt. W: width of the belt; V: speed of the belt; Q: capacity of the belt; P: power of the drive unit; XRD: Robbins crossover machine; EPB: Earth Pressure Balance; DS: double shield; SH: single shield.
*The belt used in Kargı Project has a curve radius of 1,524 m, 4 carrying side boosters 370 kW, 2 return side boosters 75 kW, and a radial stacker at the portal.

FIGURE 5.4 A typical view of the belt conveyor from Eşme high-speed tunnel, photographed by the author, S. Acun.

FIGURE 5.5 A typical view of belt conveyor from Kargı power tunnel, photograph by the author, S. Acun.

TABLE 5.2
The Characteristics of Different Sections of the Metro Line

Station	Length (m)	Gradient (%)	Curve Radius (m)
Yenimahalle–Karadeniz	1,510	1.5	800 to right, 350 to left
Karadeniz–Tekstil Kent	1,600	2.5	None
Tekstil Kent from Point A to B	550	0.0	600 to right, 414 to left
Tekstil Kent from Point B to Yüzyıl	800	−4.4	

Note: Total length of the belt conveyor from Yenimahalle to Yüzyıl Station is 4,460 m.

The maximum capacity of the belt is calculated as $Q = A \times V \times \gamma$; γ being the density of the loose material on the belt for EPB tunneling.

$$Q = 0.03 \text{ m}^2 \times 3 \text{ m/sec} \times 1.4 \text{ t/m}^3 \text{ or } Q = 0.126 \text{ t/sec or } Q = 454 \text{ t/h}$$

 d. The power of the belt drive

The total power of the belt drive is the sum of the power to drive the empty belt, the material on the belt, and the power to handle with the gradient.

The power to drive the empty belt may be calculated using Equation 5.1.

$$W_e = m_i \times (1 + 1x) \times g \times \mu e \tag{5.1}$$

In Equation 5.1, W_e is the power to drive the empty belt, m_i is the weight of one meter of belt in kg/m and empirically it may be calculated as $m_i = 60w$ as Brook (1971) suggested. However, recently there are newly designed more powerful belt conveyors in the market where m_i may equal to 70w.

l+lx is the length of the belt plus the extra length of the belt on the drum, which is 25 m in our case.

 g is the center of gravity which is 9.81 kg/cm².

 μe is the friction coefficient, and a value of 0.03 is generally used.

 For the distance between Yeni Mahalle and Karadeniz Mahallesi stations of the length 1,510 m, the power to drive an empty belt may be calculated using Equation 5.1 as below.

 $W_e = 70 \times 0.95(\text{kg/m}) \times (1{,}510 \text{ m}+25 \text{ m}) \times 9.81(\text{kg/sec}^2) \times 0.03 \times 3(\text{m/sn})$

 W_e is calculated as 90,042 W or 90 kW

The power W_m to drive the material on the belt may be calculated using Equation 5.2.

$$W_m = T \times l \times g \times \mu m \tag{5.2}$$

In Equation 5.2, T is the quarrying capacity, l is length of the belt, and µm is the friction coefficient between the material and the belt.

The power to drive the material on the belt for the distance between Yeni Mahalle and Karadeniz Mahallesi stations may be calculated as given below.

W_m = 110,000(kg/3,600 sec) × (1,510 m + 25 m) × 9.8(m/sec^2) × 0.04

W_m = 18,386 W or W_m is calculated as 18.4 kW

The power W_r to raise the material between Yeni Mahalle and Karadeniz Stations may be calculated as in Equation 5.3.

$$W_r = T \times g \times h \tag{5.3}$$

T = 110,000 (kg/3,600 sec) × 9.81(m/sec^2) × 1,510 × 1.5(m/100)

W_r = 6,789 W or 6.8 kW

The summary of the calculations for four segments of the Metro Project is given in Table 5.3

It is interesting to note that the calculated value of driving power is very close to actual values given in Table 5.1, and only 31% of the driving power is for conveying the material and 69% of the power is spent for mowing the belt itself.

e.   Calculation of driving belt tension and selection of belt material

The driving drum relies on the friction between drum and belt to give the drive to the belt. The top strand normally pulls the belt and load over all the troughing idlers, and in many cases also lift these loads up an incline. The lower strand tension depends on the belt being sufficiently tight all along its length for some tension to be remaining at the point just after the belt leaves the driving drum. Belt conveyor troughing idlers and driving drum belt tensions are illustrated in Figures 5.6 and 5.7. If the two tensions in the belt at the driving drum are P_1 and P^2, with P_1 the bigger

---

TABLE 5.3
Summary of the Calculations Made for the Four Segments of the Metro Project

Station	W_e (kW)	W_m (kW)	W_r (kW)	Total Power (kW)
Yenimahalle–Karadeniz	90.0	18.4	6.8	115.2
Karadeniz–Tekstil Kent	95.4	19.5	11.3	126.2
Tekstil Kent from Point A to B	32.8	6.7	6.7	46.2
Tekstil Kent from Point B to Yüzyıl	47.7	9.7	0.3	57.7
	265.9	54.3	25.1	545.3

Source: From the archive of the author, N. Bilgin.

Notes: 0.9 is taken as the efficiency of the driving engine; total power for four segments is 345.3/0.9 = 384 kW.

FIGURE 5.6   Belt conveyor troughing idlers.

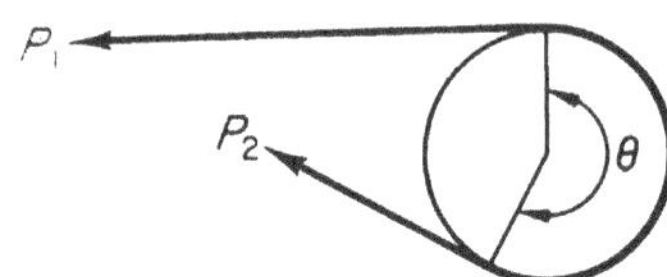

FIGURE 5.7   Driving drum belt tension.

tension in the top strand, the limiting ratio of tensions for no-slip occurs is given by Equation 5.4

$$\left(\frac{P_1}{P_2}\right) < e^{\mu\theta} = n \tag{5.4}$$

$$P_1 = \left(\frac{n}{n-1}\right) \times \left(\frac{Power}{V}\right) \tag{5.5}$$

Where μ is the coefficient of friction or coefficient of grip between the belt and drum, and θ is the angle of the wrap as in Figure 5.7.

Slip is avoided in practice by using a value of n less than that possible by assuming a low value for μ, the value of θ being fixed by the arrangement of the drum at the angle θ is about 250° for a single-drum drive, and 400–450° for a two-drum drive. The value μ changes between 0.2 and 0.35 according to the material used on belts.

The belt material is selected according to calculated values of P_1.

θ and μ are selected as 440° and 0.2.

$e^{\mu\theta} = e^{0.2 \times 440(\pi/180)}$ which is found as 4.65.

And using Equation 5.5, $P_1 = (4.65/3.65)(345.3/3)$.

The maximum tension on the belt is then obtained as $P_1 = 147$ kN

The maximum tension per unit belt width is then calculated as $P_1/W_1 = 147$ N/mm.

The strength of the belt material is selected according to the calculated maximum tension value on the belt.

The calculated values for the selection of the belt conveyor are in the same order as used for Mahmutbey Mecidiyeköy Metro Project. However, during the course of the project, it was observed that belts were breaking off in places where curvature existed, and boosters were used in the critical points to solve the problem.

5.2.4 Vertical Belts

Vertical belt conveyors can move a large amount of muck as quickly as horizontal belts. Vertical belt conveyors can bring the rock up a shaft to the surface. Whether the tunnel is 20 m or 200 m deep, vertical belt conveyors will reliably deliver the rock to the surface with minimal spillage. There are custom designs for each vertical belt conveyor specifically to suit the project, including muck transfer from the horizontal conveyor to the vertical belt conveyor; muck transfer from the vertical belt conveyor to overland or stacker conveyor; structural mounting of the conveyor in the shaft; muck recycling chutes to minimize spillage; with safety interlocks and worker protection.

To enable 1,600 t of tunnel muck to be removed per hour from the Grand Paris Express tunneling project, H&E Logistics of Bochum, Germany has commissioned two Dos Santos International (DSI) sandwich belts, high-angle conveyors. Each conveyor elevated 800 t of muck per hour at a 90° angle (*Tunnel Talk*, High angle muck removal for Paris project, 28 Sep 2017). https://dossantosintl.com/get-the-muck-out-with-sandwich-belt-high-angle-conveyors/

The vertical belt conveyor with DSI sandwich belts used in the Grand Paris Express tunneling project is seen in Figure 5.8 and its characteristics are given in Table 5.4.

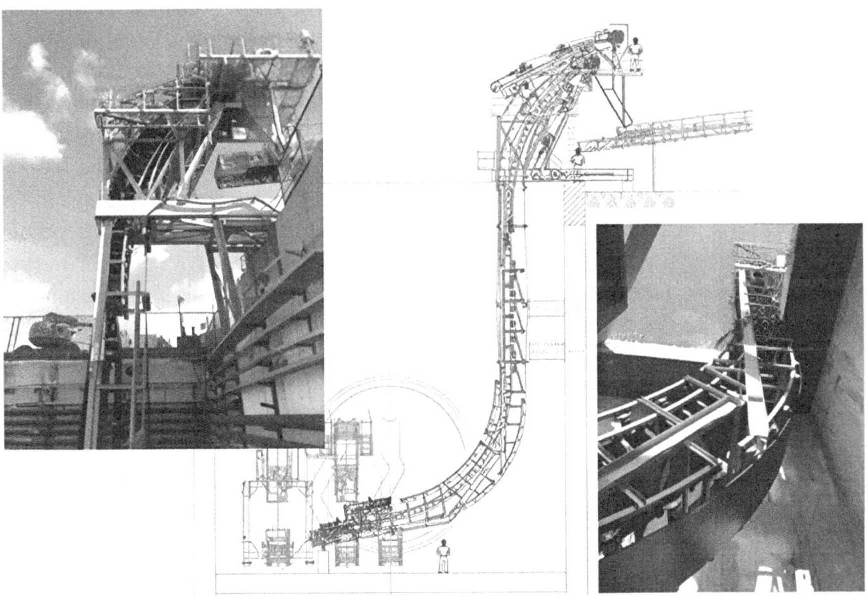

FIGURE 5.8 Vertical belt conveyor with DSI sandwich belts used in the Grand Paris Express tunneling project, for H&E Logistics of Bochum. (*Tunnel Talk*, High angle muck removal for Paris project, 28 September 2017.)

TABLE 5.4
DSIGPS Sandwich Conveyor for H+E Logistic at
Paris Metro Project with courtesy of Herrenknecht

Material	Tunnel Muck
Density	1.6 t/m^3
Size	150 mm minus
Conveying rate	800 t/h
Conveying angle	90°
Belt width	1.4 m
Belt speed	3 m/s
Lift	26.3 m
Length	35.2 m
Snake drives, top belt	75 kW
Snakes drives, bottom belt	75 kw

5.3   MUCK TRANSPORT WITH LOCOMOTIVES

Finding the size and specifications for a locomotive may require the consideration of many factors. The first and most important step is to decide on the required locomotive weight necessary to develop the tractive effort to move the trainload. The maximum grade and the load are the inputs for this calculation. A locomotive with steel tires on clean dry steel rails can produce a running tractive effort of 26% of its weight. A locomotive should be able to produce about 15% more tractive effort than required by the trailing load to provide a reasonable acceleration rate (Buckeridge et al. 1982).

Locomotives are limited to shallow gradients and the frictional grip between the driving wheels and the rails also limits the size of the train which can be hauled. A torque is applied by the engine, and slip is prevented in normal circumstances by the friction force at the point of contact between the wheel and the rail. The friction force is limited to the mass of the locomotive and to the coefficient of friction. The friction force provides the forward force pulling the locomotive and train and is termed the tractive effort. The tractive effort has to overcome various resistances to motion, which are, friction resistance, gradient resistance, and inertia resistance (Brook, 1971).

Diesel locomotives are used in tunneling operations since in battery locomotives batteries need to be charged very often complicating the system. The characteristic of the diesel engine is to give a steady torque over a wide range of speeds. The value of the torque depends on the amount of fuel injected per engine cycle, which in turn is controlled by the driver's throttle. The effective range of speed is not enough to allow large loads to be started and to give fast haulage speeds, and some form of speed gearing is required in transmission (Brook, 1971).

Robust, functional, and maintenance-friendly hydrodynamic locomotive designs exist in the market. These types are available in the weight range from 18 to 45 t with corresponding engine power from 125 to 360 kW in gauges up to 1,000 mm, suitable for tunnel diameters from 5 to 15 m and gradients up to 3%. A practical alternative to double traction with two locomotives, hydrostatic-driven platform cars provide a significant increase in tractive power and are well suited for transporting overburden buckets, mortar mixers, or other loads.

California switch is indispensable to ease the passage of the train. A portable or fixed platform rides on a track in a tunnel and is used for passing cars and trains. It has space for two or more trucks, crossovers, and switches, and has sliding and tapered end rails that ride on main-line rails. Optimizing the transport systems by increasing the number of trains and adding California switch increases TBM utilization time. One of the biggest problems is getting people and materials in and out from the tunnel and the problem may be reduced by using California switches (Ponnuswamy and Victor, 2016). When using muck cars in large-diameter tunnels, double-track backup systems can be employed to make maximum production. By installing a California switch behind the backup, empty muck cars can be kept near the loading point while others are being filled. This minimizes delays between trains and optimizes advance rates. A typical California Switch is seen in Figure 5.9.

5.3.1 Some Examples of Using Locomotives in Tunnels from Turkey

5.3.1.1 Kozyatağı Kadıköy Metro Tunnels

Two twin tunnels were excavated in Istanbul between Kozyatağı and Kadıköy with two TBMs of 6.57 km in diameter.

Due to a high degree of inclination of % 3, a tandem locomotive system was selected. Five cars of 12 m^3 (with 20 t, with muck in), 25 t of weight were used.

Fixed California switches were preferred due to small distances between stations.

5.3.1.2 Yamanlı Energy Tunnel

Energy Tunnel of 6,618 m in length of Yamanli II HEPP Project was constructed with a double shield TBM of 4.305 mm excavation diameter.

Schoema locomotives of 20 t and 100 kW of power were used with muck cars of 8 m^3 with self-discharger and one fixed and one moving California switches were used.

Rolling stock provided by MSD for the construction of a 9.3-km water tunnel for the hydroelectric power plant Yamanli II, Turkey is as below (Figure 5.10):

- 8 segment cars
- 8 flat cars of 8 m
- 4 pea gravel cars of 3 m³ (with conveyor)
- 16 muck cars of 8 m³ – self discharger
- 2 discharging devices
- 2 riders for 10 persons each – Uniblock
- 1 California switch

5.3.1.3 Afşar Hadimi Water Tunnel/Turkey

The tunnel is planned to deviate water from Afsar Hadimi Dam to Bagbasi Dam reservoir. Then, this tunnel will transfer the water via Mavi Tunnel to Konya-Cumra basin.

FIGURE 5.9   A typical California Switch.

An EPB-TBM having an excavation diameter of 5.25 m was used in the project. Tunnel length was 11,500 m. A locomotive of 155 kW of 25 t in weight was recommended for the project.

5.4   TRACKLESS TRANSPORT (MSVS) FOR MATERIAL TRANSPORT

MSVs are designed to carry heavy loads up to 200 t in narrow environments. They may work in inclines and declines up to 20%. In tunnels they are used on the job site to transport all operating materials needed for work in the tunnel, like precast

FIGURE 5.10 Muck transport system used in Yamanlı Energy Tunnel. Courtesy of MSD Company.

TABLE 5.5
Characteristics of Multi-service Vehicles to Supply TBM Tunneling, by TMS

Parameter	MSV Single	MSV Double	MSV Triple
Load capacity (t)	10–65	40–140	120–200
Width (m)	1.0–1.2–1.5–1.7–1.8	1.5–1.7–1.8	1.9
Number of wheels	4, 6 or 8	8–20	24
Drive wheels	4, 6 or 8	8–20	12–24
Drive (kW)	100–150–200–315	200–315–400	315–400
Maximum speed (km/h)	25,	20	16
Maximum incline (%)	25	20	15
Steering system	Electronic	Electronic	Electronic

Source: From the catalogue of TMS.

segments, grout tanks, material platforms, and personnel modules. Unlike rolling stock, MSVs do not require rails, saving a huge amount of time and money in the setup and cleaning of rails on the job site. They have extremely high-performing braking distances. They also have the ability to move to both sites.

Some characteristics of MSVs manufactured by TMS to supply TBM tunneling are given in Table 5.5 and a typical photograph of an MSV carrying out precast segments are given in Figure 5.11.

FIGURE 5.11   An MSV carrying precast segments.

5.5   RISK MANAGEMENT PROCESS FOR ROLLING STOCK-LOCOMOTIVES, RAIL CARS, AND CONVEYORS IN TUNNELS

The following risk management processes for rolling stock-locomotives, rail cars, and conveyors in tunnels are a summary obtained from the Code of Practice for tunnels under construction which is released by the New South Wales Government (2006).

5.5.1   RISK MANAGEMENT FOR ROLLING STOCK-LOCOMOTIVES, RAIL CARS IN LONG TUNNELS

As well as the general issues that apply to all plants, the risk management process for rolling stock should consider the following issues:

- Maximum grade
- Fail-safe brakes with dead-man switch control
- Speed limiters, control of speed which may occur due to load variation
- Couplings and safety chains
- Signaling systems
- The rail track, switchgear, passing zones
- Pedestrian access
- Appropriate rolling stock
- Derailments, including prevention measures, recovery systems, and equipment
- Tipping systems
- Buffer stops.

5.5.2 RISK MANAGEMENT PROCESS FOR BELT CONVEYORS

Conveyors should conform to AS 1755: Conveyors – safety requirements, which sets out the minimum safety requirements for the design, installation, guarding, use, inspection, and maintenance of conveyors and conveyor systems. The risk management process should identify and eliminate (or control) the hazards and risks associated with conveyors. For tunneling, the following controls should be considered:

- Isolating conveyors from normal work areas to prevent entanglement of limbs or body
- Preventing personnel from riding on conveyors
- Using fire resistant or fire resistant anti-static conveyor belting
- Providing fire extinguishers
- Preventing oversize material from entering the conveyor system
- Reducing spillage from overloaded conveyors and regulating the conveyor's feed rate and belt speed
- Suppressing dust
- Implementing power isolation procedures to allow for maintenance, spillage clean-up, and clearing the rollers
- Implementing maintenance systems that consider the increased risk, fire, or falling objects, posed by failed bearings on idlers and drums
- Implementing procedures after shutdown and maintenance.

5.6 CONCLUDING REMARKS

In a tunneling operation, the muck transportation with a belt conveyor or locomotives affects the overall performance. The conveyor belt shows an average increase of 3% in TBM utilization as the length of the tunnel increases in hard rock. However, in some cases, this average may go up to 5%. The length of the tunnel also plays an important role in determining machine utilization. In TBM drives, if the transportation distance is longer than 1.5 km, the stoppages in rail transport are much greater compared to belt conveyors. It is always the contractor's decision, but generally, if the tunnel is longer than 1.5 km in length, conveyors are used. If it is shorter, then locomotive haulage is usually preferred. In practice, belt conveyors smaller than 60 cm wide are not used. On the other hand, because of its lower investment cost compared to belt conveying, some contractors prefer locomotive transport.

In long tunnels or in tunnels having curvatures, boosters are used to reduce the overall belt tension. In this way, the excavated material can be transported safely over long distances also around the tightest curves. The location of the booster drives is critical since the location directly affects the maximum tension developed in the belt.

Vertical belt conveyors are also used in shafts and they can move a large amount of muck as quickly as horizontal belts. Vertical belt conveyors can bring the rock up a shaft to the surface. Depending on the depth of the shaft ranging between 20 m and 200 m, vertical belt conveyors may reliably deliver the rock to the surface with minimal spillage.

The suppliers usually make calculations for muck transportation systems, belt conveyors or locomotive transport on a given project and make some suggestions. However, the designer must also make some primary calculations, which will allow him to suggest some recommendations for a proper muck transport system and the tunnel engineer must understand all the calculation steps to have primary data for the management of the project. When using muck cars in large-diameter tunnels, double-track backup systems make sure of maximum working efficiency. By installing a California switch behind the backup, empty cars can be kept near the loading point while others are being filled. This minimizes delays between trains and optimizes operational parameters.

MSVs are designed to carry heavy loads up to 200 t in narrow environments. They are faster than locomotives. They may work in gradients up and down to 20% in tunnels. They are used in job sites to transport all operating materials needed for work in the tunnel, like precast segments, grout tanks, material platforms, personnel, and modules. Unlike rolling stock, MSVs do not require rails, saving a huge amount of time and money on the setup and cleaning of rails on the job site. They have the ability to move to both sites.

The risk management process for rolling stock locomotives, rail cars, belt conveyors, and MSVs in tunnels is of prime importance, and the code of practices released by some government organizations must be strictly followed. The mechanical engineer is responsible for the maintenance and safety required by the regulations and his staff should be supervised carefully and effectively regarding the installation, examination, and maintenance of the related equipment.

REFERENCES

Aşık, E., 1988. *Belt Conveyors Principles of Construction and Calculation* (In Turkish), Chamber of Mechanical Engineers, 98, p. 333.

Bilgin, N., 1977. The common aspects of mining and tunnel engineering and some recommendations, in *Mining Congress Organized by Chamber of Mining Engineers*, p. 8, Ankara, Turkey.

Bilgin, N., Acun, S., Ates, U., Murtaza, M., Çelik, Y., 2017. The factors affecting the performance of three different TBMs in complex geology in Istanbul, *Proceedings of the World Tunnel Congress 2017 – Surface Challenges – Underground Solutions*. Bergen, Norwege.

Bringiotti, M., Bringiotti, G., Aebersold, W., Rufer, P., 2019. Moving material with belt conveyors in urban environment and for long tunnel construction in Italy: Metro Catania, Scilla's shaft, A1 highway and Brenner Base Tunnel, in *Tunnels and Underground Cities: Engineering and Innovation Meet Archaeology, Architecture and Art*, D. Peila, G. Viggiani, & T. Celestino, (eds), Taylor & Francis Group, London, p. 436.

Brook, N., 1971. *Mechanics of Bulk Materials Handling*, Butterwords, London, p. 166.

Buckeridge, R., Crey, W.T., Graham, A.H., Martino, S.M., Resee, C.D., 1982. *Rail Haulage Systems*. Underground Mining Methods Handbook, AIMM, p. 1227.

CEMA (Conveyor Equipment Manufacturers Association) 2020. *Belt Conveyors for Bulk Materials*, 7th Ed. p. 815.

Farrokh, E., Rostami, J.I., Askilsrud, O., 2013. Downtime analysis of hard rock TBM case histories. A methodology of using past experiences in the performance prediction of a TBM in complex geology and risk analysis, in *World Tunnel Congress 2013 Geneva Underground – the Way to the Future!*, G. Anagnostou & H. Ehrbar (eds), Taylor & Francis/Balkema, Leiden, the Netherlands.

Fossati, D.A., Jakobsen, P.D., Multan, M.A, 2017. Tunnelling experiences on using conveyor belt for mucking at the world-longest and deepest subsea road tunnel, in *Proceedings of the World Tunnel Congress 2017 – Surface Challenges – Underground Solutions*. Bergen, Norway.

Khetwal, A., Rostami, J., Priscilla, N.Ç., 2021. Effect of muck transportation system on overall TBM performance and downtimes. *Conference*: *Rapid Excavation & Tunneling Conference* 2021, pp. 476–483, USA.

Komatsu-Joy, 2021, Tunnelling conveyor systems, brochure downloaded in 2021 from a web page, https://mining.komatsu/docs/default-source/product-documents/surface/crushing-and-conveying-equipment/tunnelling-conveyor-systems-brochure.pdf?sfvrsn=5073 0a6b_26.

Komatsu-Joy, 2022, Tunnelling Conveyor Systems - Komatsu Mining Corp. https://mining. komatsu › default-source › surface.

Namlı, M., Cakmak, O., Pakis, I.H., Tuysuz, L., Talu, D., Dumlu, M., Balci, C., Copur, H., Bilgin, N., 2013. A methodology of using past experiences in the performance prediction of a TBM in complex geology and risk analysis, in *World Tunnel Congress 2013 Geneva Underground – The Way to the Future!*, G. Anagnostou & H. Ehrbar (eds), Taylor & Francis/Balkema, Leiden, the Netherlands, p. 12.

Namlı, M., Bilgin, N., 2017. A model to predict daily advance rates of EPB-TBMs in complex geology in Istanbul. *Tunnelling and Underground Space Technology*, 62(2017), pp. 43–52.

Ponnuswamy, S., Victor, J., 2016. *Transportation Tunnels*, CRC Press/Balkema, Rotterdam/ Brookfield, p. 355.

Rostami, J., 2016. Performance prediction of hard rock tunnel boring machines (TBMs) in difficult ground. *Tunnelling and Underground Space Technology*, 57, pp. 173–183.

6 Cost Management

6.1 INTRODUCTION

The accurate cost estimation in tunneling projects is the key factor for the success of the project. Especially in the preliminary design stages, such cost data plays an important role in project development, affecting the major decisions to be taken. The cost of tunneling projects is greatly influenced by a number of parameters: equipment, labor, and consumables, with the most influential being the geological and geotechnical conditions encountered. That is why in a cost study analysis, the benchmarking and comparison of projects on a country basis is important. Reilly (2005) found that specific, relevant project cost information was usually unobtainable. Little objective history can be found, including findings that would support recommendations for improvement. Because of the difficulty in obtaining the data, firm conclusions could not be reliably drawn, and in the past, it was reported that there were significant cost and schedule overruns due to poor management resulting in at least 30%, and possibly more than 50%, of the projects. Examples include the Jubilee Line Transit Project in London which is 2 years late and £1.4 billion (67%) over budget, the Channel Tunnel with £3.7 billion (80%) over budget, Denmark's Great Belt Link with 54% over budget, etc. (Reilly 2005). This is why comprehensive and accurate data on tunnel construction costs should be made publicly available to the tunnel engineering community in an attempt to form a worldwide project database, which could be used as a basis for a sound project cost analysis, leading to proper cost management (Benardos et al. 2013).

Project cost management includes activities and tools to help to complete the tunneling project within the approved budget. It is also a process of estimating, budgeting, and controlling costs throughout the project life cycle, with the aim of keeping expenditures within the approved budget. Projected costs are calculated during the planning phase of a project and must be approved before work begins. As the project plan is executed, expenses are documented so things stay within the cost management plan. Once the project is completed, predicted costs and actual costs should be compared, providing benchmarks for future tunnel cost management plans and tunneling project budgets.

Underground projects and mostly public transportation projects require major financial investments. The construction cost estimates need to be combined with risk

DOI: 10.1201/9781003358978-6

analysis for the success of the project. An unrealistic cost estimate may cause project failure in the early phase. The public sector faces major problems with financing big projects due to budget and time constraints. In order to realize successful project implementation, there is a need to manage the tunneling project with the utmost care.

This chapter is written with the objective of summarizing basic factors governing tunneling costs in order to give a sound basis for cost management.

6.2 BASIC PARAMETERS IN TUNNEL COST MANAGEMENT PLAN IN MECHANIZED TUNNELING

The following items should be included in a cost management plan.

6.2.1 SPECIFICATIONS OF THE TUNNELING PROJECT OR PLANNING RESOURCES

The tunneling projects are planned either by government authorities such as metro projects, infrastructure projects, or by private sectors such as tunnels for hydroelectric power plants. In each case, the length and diameter of tunnels are specified according to the needs of the projects. Extensive geological and geotechnical studies will determine the tunneling method to be used: conventional or mechanized. Each method has different specifications for tunnel support, muck transport, ventilation, environmental, safety issues, etc. All the specifications determined during project planning will determine the tunnel cost.

6.2.2 TUNNEL COST ESTIMATING

Costing is the process in which a contractor attempts to estimate the costs involved in different tunneling activities. Costing requires the use of historical information, which is concerned with the past costs incurred by the tunnel contractors, and this information is used to predict the future cost structure of the contractor. Costing is indispensable for a contractor as it allows the company to evaluate its current cost levels, estimate the costs to be incurred in the future, and arrange to reduce those cost levels. The items considered in a tunnel cost estimate are summarized in the following.

6.2.2.1 Purchase of the Major Tunneling Equipment

In the case of mechanized tunneling, the major equipment to be considered may be TBM (hard rock TBM/single shield/double shield, EPB, slurry, etc.), rodheaders or impact hammers to excavate cross passages, etc.

6.2.2.2 The Cost of the Items Supporting the Tunnel Drives

Tunneling support costs include those on the contractor's site establishment, project management, and supervision. These costs are split into *fixed and time-related* to allow a more representative estimate to be prepared.

Fixed costs elements are those not influenced by time or tunnel length and include establishing a secure site compound that contains serviced site offices, workshops, testing laboratories, stores, and places for the storage of segments. Gantry cranes, high voltage electrical supply to the TBMs, a railway system to move people and materials between the portal and the TBM, equipment to transport excavated material from the TBM face to the portal, and tunnel ventilation equipment are among the fixed costs. For a slurry TBM, a slurry treatment plant represents a substantial fixed cost.

Time-related costs include labor costs for management, supervision, general site-based labor, hired plant and equipment servicing the site, security, cleaning, and maintenance costs.

6.2.2.3   Tunnel Construction

It comprises the excavation of the material from the face of the tunnel and its transport along the tunnel to the tunnel construction site. The cost of the segments and their transport and installation in the tunnel, taking out the temporary construction equipment upon the completion of tunnel boring, the installation inside the completed tunnel of a concrete base incorporating drainage pipes, a concrete evacuation walkway, and a concrete maintenance walkway are important items of the tunnel construction. Cross passages to link twin tunnels must also be constructed every 300–400 m of the tunnel length. In the event of an emergency in one tunnel rendering the train unable to continue to an open section of the route, the cross passages allow passengers to evacuate from the affected tunnel to the unaffected tunnel.

6.2.2.4   Disposal of Excavated Material

Excavated material that arises from bored tunneling, either from a slurry or an EPB-TBM, is generally too wet to be used for engineering purposes on the route. It is therefore required to be transported out of the construction site to a damping place where environmental restrictions should be followed with utmost care.

6.2.2.5   Portal Construction

Portal construction is widely discussed in Chapter 4.

6.2.2.6   Shafts Construction

For tunnels of sufficient length, ventilation shafts are required to be constructed that connect both of the tunnels to the surface. The shafts provide ventilation and intervention access for emergency services and during operation and maintenance.

6.2.2.7   Mechanical and Electrical Systems

They are required in the tunnels, cross passages, and ventilation shafts to light, drain, ventilate, and cool the tunnels. Fire and life safety equipment is also required comprising fire hydrants, signage, and communication systems.

6.2.3   BUDGETING OF TUNNELING COSTS

Budgeting is concerned with planning for the future and involves the tunneling contractor, making a plan regarding the costs to be incurred for each tunneling activity, and ensuring that payments are made out of the budget allocated in the plan.

Budgeting allows a contractor to keep its costs efficiently at the planned levels and results in less overspending. It is an essential part of controlling company finances and will help the contractor to be prepared for unexpected events, avoid a financial crisis, make better returns from funds used, and be an essential part of the planning process.

6.2.4 THE CONTROL OF TUNNELING COST

Cost control is achieved by collecting actual tunneling cost data for each area of tunnel activity, comparing actual cost values with anticipated cost values, and forwarding the prompt report to top management highlighting the deviations from anticipated values for immediate corrective action. Further, cost control needs executive action and urges actual costs to conform to planned costs.

6.3 COST DISTRIBUTION IN MECHANIZED TUNNELING

Infrastructure projects such as tunneling projects are of prime importance within the tunneling community. They may be executed either by conventional or mostly by mechanized tunneling methods. Understanding the cost-affecting items within the total cost plays an important role in the management of the project. Typical cost distribution in the metro in urban areas (soft ground) is summarized as defined in Table 6.1 (Wagner 2004). However, it is important to note that the cost of tunnels having the same diameter and the same length may be different, whether it is a highway tunnel, subway tunnel, water tunnel, or wastewater tunnel as Rostami et al. (2013) pointed out in a research study covering 270 projects.

In international basis, it should be emphasized that the tunneling cost is mainly related to the local facilities of each country, labor, material, etc. Figure 6.1 gives the cost distribution in Istanbul New Airport Metro Project, Gayrettepe-Airport Line. It

TABLE 6.1

Typical Cost Distribution in Mechanized and Conventional Tunneling in Urban Areas (Soft Ground)

Item (percentage of the costs within the total cost)	Mechanized Tunneling	Conventional Tunneling
Equipment (%)	24	14
Material (%)	23	29
Personnel (%)	19	21
Overhead (%)	11	11
Unexpected expenditure (%)	8	5
Transport (%)	6	14
Site mobilization/clearance (%)	5	12
Additional parts (%)	4	7

Source: Wagner (2004).

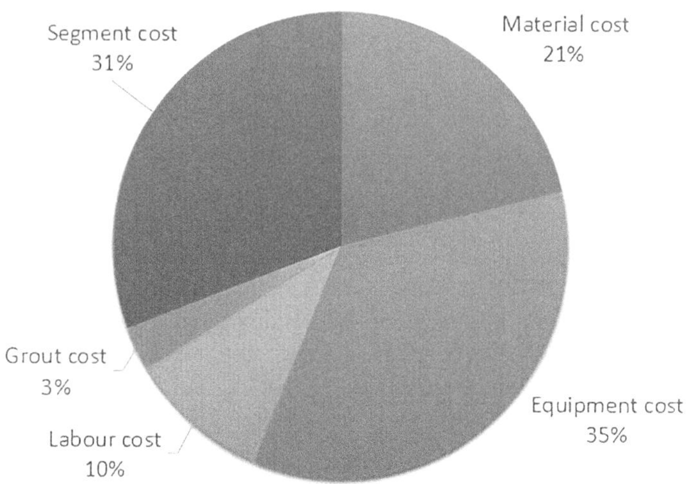

FIGURE 6.1 Cost distribution in Istanbul New Airport Metro Project, Gayrettepe-Airport Line. (From the archive of the author, N. Acun.)

is interesting to note that the cost of the equipment and precast segment rings are the major cost items being 35% and 31% within the total costs of labor, material, grout, equipment, and precast segments.

Munfah and Nicolas (2020) lately examined the cost of the US tunneling projects emphasized on the following items.

The costs in the US tunneling projects are generally as follows:

- 55% of the total cost is construction cost.
- 35% of the total cost is soft costs, which include owner cost, preconstruction costs including environmental impact, feasibility studies, program management consultant, design consultant, construction management, permits, insurance, finance, bonding, etc.
- 10% of the total cost is third-party costs, which include utility diversions, remedial works, and stakeholders' commitments.

Evaluating the construction cost alone, the approximate breakdown is defined as follows:

- Labor cost is 40% to 50% of the construction cost.
- Contractor construction equipment TBM, etc., is 18% to 20% of the construction cost.
- Permanent material is 15% to 18% of the construction cost.
- Construction material, temporary works, consumables, etc. costs are 10% to 12% of the construction cost.

6.4 THE COMPARISON OF THE TUNNELING COSTS IN DIFFERENT COUNTRIES

The comparison and interpretations of tunneling costs in different countries may serve better to explain the factors affecting the tunneling costs. Table 6.2 provides a comparison of major projects from around the world; as can be seen, the cost of tunneling varies greatly around the world.

As explained earlier, labor cost is 40% to 50% of the construction cost.

Mutaf and Nicolas (2020) explained the difference in tunneling costs emphasizing on labor costs and environmental restrictions as explained in the following paragraph.

The comparable projects in Europe are generally constructed at a lower cost than similar projects in the United States. Australian project costs are closer to US projects than any other country; this is believed to be due to the significant environmental requirements in Australia. Labor cost and construction schedule are the most important factors affecting construction cost. Labor cost is often driven by labor union rules, which vary significantly among states and cities. One of the highest labor costs of tunnel construction workers is the Sandhogs in New York, which can be as high as $110/h, and on an overtime basis, it can reach over two to three times this value. Their rates are higher than other tunnel workers in the

TABLE 6.2
The Comparison of the Tunnel Costs in Different Countries Based on the Project Overall Cost

Project	Year	Cost (USA Dollars 10^6 /km)
Tokyo metro	2008	250.0
Berlin metro	2009	204.3
Barcelona metro	2009	151.9
No.7 Subway extension, NY	2015	937.7
U-Link, Seattle, USA	2016	375.0
Second Ave. Subway, NY	2018	1,562.5
East side access, NY	Planned 2022	1,125.0–1875.0
Purple line (1 and 2), Los Angeles	Planned 2025	500.0
Regional connector, Los Angeles	Planned 2021	575.0
Central subway, San Francisco	Planned 2021	575.0
Cross rail, London	Planned 2021	312.5
Melbourne metro	Planned 2026	468.8
Sydney metro city	Planned 2024	468.8
Grand Paris metro, different segments	Planned 2030	193.8–281.3
Doha metro, different segments	Planned 2026	125.0–250.0
Seoul Shinbundang line	2011	87.5
Hangzhou metro (China)	2012	143.8
Mumbai, Delhi, Kolkata metro, India	Varies	62.5–93.8

Source: Mutaf and Nicolas (2020).

Note: The tunnels are typically twin-bore with inside diameters in the range (6–6.7 m).

United States and significantly higher than European or Asian workers' rates. In addition, the number of workers assigned to the tunnel in New York is significantly more than in other parts of the country and as much as four times more than tunnel workers assigned to comparable projects in Europe. Whereas in Germany (one of the high labor-rated countries in Europe) the comparable labor rate is about $30/ h. In other European countries such as Spain, Portugal, Italy, Poland, etc., labor rates are even lower.

6.5 THE EFFECT OF TUNNEL DIAMETER AND TUNNEL LENGTH ON TUNNELING COST

Figures 6.2 and 6.3 represent the effects of the tunnel outside diameter and tunnel length on unit costs. The figures are taken from a report jointly published by HM Treasury and Infrastructure UK. The source of the figures was obtained from UK cost questionnaires and the British Tunneling Society. The tunnels are from the rail, highway, water, and power sectors, and from Norway, Spain, Netherlands, Austria, Portugal, Germany, Switzerland, France, Greece, and Luxembourg. The data points represent the outrun cost of each tunneling contract, including portals and shafts, divided by the total length of the tunnel drives. The main conclusions obtained from these figures are that the cost increases with tunnel diameter and decreases with tunnel length. The main reason for the scatter in the graphs is that tunneling cost depends mainly on ground conditions, tunneling method, and lining type. However, it should be also mentioned that although the tunnel cost increases with tunnel diameters, there is a benefit of using large-diameter tunnels. With increases in tunnel diameter, innovative configurations of corridors within the tunnels have been developed to optimize the usage of underground space. These include double-stacked traffic decks,

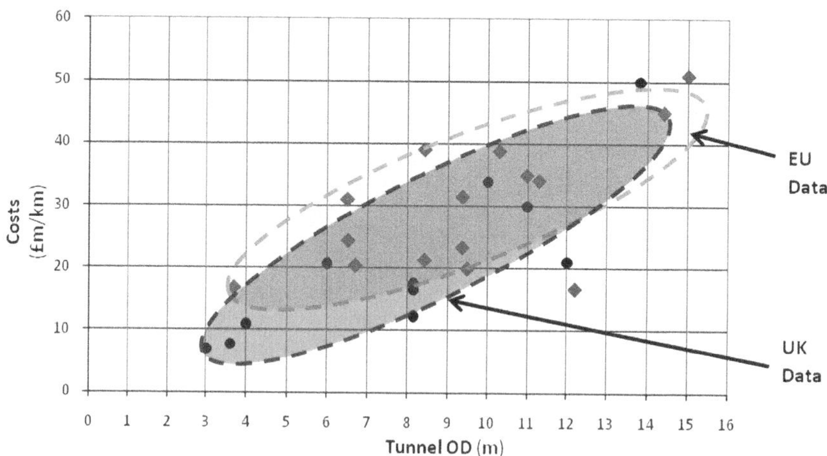

FIGURE 6.2   The effects of the tunnel outside diameter on unit costs. (HM Treasury [2010] based on the source taken from the UK cost questionnaires and British Tunnelling Society.)

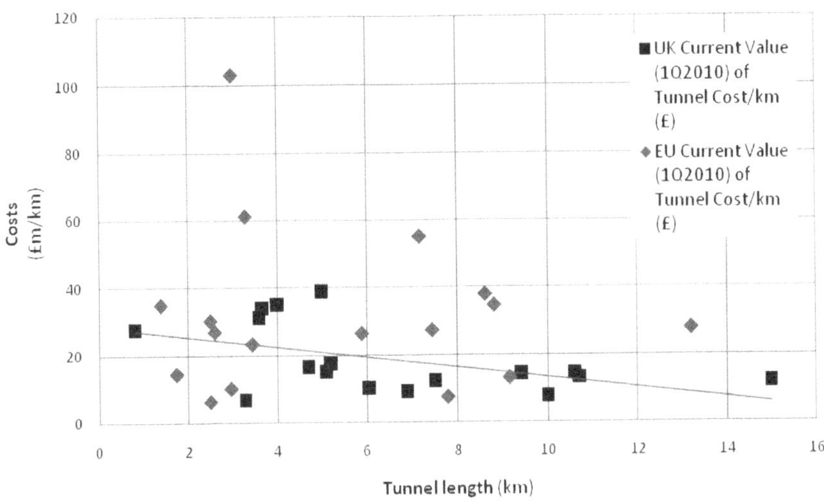

FIGURE 6.3   The effects of the tunnel length on unit costs. (HM Treasury [2010] based on the source taken from UK cost questionnaires and British Tunnelling Society.)

separation of vehicle traffic from heavy goods vehicles, and in the case of the SMART tunnel in Malaysia, the use of a part of the tunnel as storm-water overflow tunnel (Tunnel talk 2015).

6.6   TUNNEL COST DISTRIBUTION IN A RECENT TUNNELING PROJECT

Typical cost distribution of tunnels in the High-Speed 2 (HS2) project is given below based on a report titled "A Guide to Tunnelling Costs" (House of Commons 2005). HS2 is a new high-speed railway linking up London, the Midlands, the North, and Scotland serving over 25 stations. A twin-bore tunnel planned as two parallel tunnels, each containing a single rail track, will be constructed using slurry TBMs. The first of the slurry TBMs of 8.8 internal diameters, Florence, was launched in May 2021 and will dig the 16 km long Chiltern tunnels for the next 3 years. Cecilia, the second of the slurry TBMs, is also launched parallel to the first TBM. The portal of the tunnels is seen in Figure 6.4.

This report on the tunnel cost estimate provides a general description of the principal cost elements for bored tunnels constructed either using an Earth Pressure Balance Machine or a slurry TBM. A slurry TBM is selected to be used through the Chalk of the Chilterns. HS2's designs and associated cost estimates were at an early stage of development and were therefore based on many assumptions. For HS2, the bored tunnels of 8.8 m internal diameter are spaced approximately 20 m apart (centerline to center line), lined with 400 mm thick precast concrete segments. The expected advance rate is predicted as 80 m per week.

FIGURE 6.4 The portal of the High Speed 2 tunnel project. The photograph is from hs2enquiries@hs2.org.uk, October 2021.

The estimates in the guide include the following components:

Purchase of specialist plant; construction site establishment, maintenance, and removal on completion; contractor's project management, design, controls, supervision, and administration staff; excavation and construction of the tunnels and associated structures; supply, installation, testing, and commissioning of the mechanical and electrical systems; contractor's overheads and profit.

The estimates in the guide exclude the following components:

Land and property; railway systems, e.g., track, signaling, telecommunications, and traction power systems; employer's corporate, project management, and design teams; allowances for employer's opportunities and risks, e.g., contingency; cost escalation above the base date of second quarter 2011, e.g., inflation; operating and maintaining the assets following commencement of passenger.

Portals are approximately 30 m wide structures that allow trains to travel between ground level and the bored tunnel headwalls (and vice versa), at gradients no steeper than 3%. The depth of the portal (rail level) at the tunnel's headwall must be at least twice the diameter of the tunnel, or about 20 m, below ground level in order to safely commence bored tunneling. In flat topography, portal structures may be approximately 700 m long.

For the purposes of this guide, ventilation shafts are assumed to be required typically every 3 km. Vent shafts have an approximate cross section on the plan of 45 m × 25 m (or equivalent area). Ventilation shaft depending on location is calculated to be £10,000,000 to £30,000,000 each.

TABLE 6.3
The Summary of Tunnel Cost Estimations

This example assumes a 7 km tunnel is constructed by a slurry TBM in a rural land hilly topography

Tunnel duration: 7 km of tunnel with an advance rate of 80 m/weeks – 87.5 weeks

7 km of clear out with a productivity of 400 m/weeks – 17.5 weeks

Total duration 105 weeks

Item	Description	Quantity	Unit	Value (£)	Total Cost (£)
Purchase of TBM	Slurry TBMs	2	Nr	16,000,000	32,000,000
Tunneling support costs	Fixed costs (slurry TBM)	1	Sum	45,000,000	45,000,000
	Time-related costs	105	Weeks	1,100,000	115,500,000
Tunnel construction	Twin tunnels (slurry TBM)	7,000	m	25,000	175,000,000
Disposal of the muck	To a commercial tip	7,000	m	4,500	31,500,000
Tunnel portals	Assumed hill topography	2	Nr	20,000,000	40,000,000
Tunnel shafts	Ventilation shaft in a rural location	2	Nr	12,000,000	24,000,000
Mechanical and electrical systems	In tunnels	7,000	M	4,000	28,000,000
Total cost for the example tunneled section of the route					491,000,000
Total costs except mechanical and electrical systems					463,000,000
Total length of single tunnel (km)					14
Civil engineering cost per single tunnel (km)					33,071,429

Source: House of Commons (2005).

Notes: The costs are for year 2011.

This example assumes a 7 km tunnel is constructed by a slurry TBM in a rural land hilly topography.

Tunnel duration: 7 km of tunnel with an advance rate of 80 m/weeks – 87.5 weeks.

7 km of clear out with a productivity of 400 m/weeks – 17.5 weeks.

Total duration 105 weeks.

6.7   CONCLUDING REMARKS

Tunnel cost management is a very important one among all the tunnel project management. There are several activities in a tunnel project, several project resources work on project activities, and various materials, tools, and equipment are used to complete the tunneling project. All these require the specifications of the tunneling project, cost estimate, cost budgeting, and cost control. Basic parameters in the tunnel cost management plan and the parameters controlling the tunnel cost are discussed in this chapter. The parameters include purchasing the major tunneling equipment, the cost of the items supporting the tunnel drives (fixed and time-related), tunnel construction, disposal of excavated material, mechanical and electrical systems portal, and shaft construction. An example to the cost of a high-speed tunnel, (HS29) the UK is given at the end of the chapter (Table 6.3). The tunneling cost dependent on tunnel diameter, tunnel length, labor cost, and environmental restrictions in international projects is also discussed.

REFERENCES

Benardos, A., Paraskevopoulou, C., Diederichs, M., 2013. Assessing and benchmarking the construction cost of tunnels, *GéoMontréal 2013, the 66th Canadian Geotechnical Conference and the 11th Joint CGS/IAH-CNC Groundwater Conference.*

HM Treasury, 2010. Infrastructure cost review: Technical Report, obtainable on web. https://assets.publishing.service.gov.uk/government/uploads/system/uploads/attachment_data/file/192589/cost_study_technicalnote211210.pdf uploaded June 2022.

House of Commons, 2005. A report on "Guide of Tunnelling Costs for HS2, obtainable on web. https://assets.publishing.service.gov.uk/government/uploads/system/uploads/attachment_data/file/434516/HS2_Guide_to_Tunnelling_Costs.pdf   https://assets.publishing.service.gov.uk/government/uploads/system/uploads/attachment_data/file/434516/HS2_Guide_to_Tunnelling_Costs.pdf

Munfah, N., Nicolas, P., 2020. Why tunnels in the US cost much more than anywhere else in the world, *TBM Tunnel Business Magazine.*

Relly, J., 2005. Cost estimating and risk-management for underground projects, in *World Tunnelling Congress, Istanbul, Underground Space Use: Analysis of the Past and Lessons for the Future*, Y. Erdem & T. Solak (eds), Taylor & Francis Group, London.

Rostami, J., Sepehrmanesh, M., Gharahbagh, E.A., Mojtabai, N.A., 2013. Planning level tunnel cost estimation based on a statistical analysis of historical data. *Tunnelling and Underground Space Technology,* 33, pp. 22–33.

Tunnel talk, April 2015. Cost benefits of large-diameter bored tunnels. www.tunnels.ai/cost

Wagner, H., 2004. The governance of cost in tunnel design and construction, *1st International South American Tunnelling Symposium*, p. 6.

7 Logistics Management of the Consumables in a TBM Tunneling

7.1 INTRODUCTION

The success of tunnel drives with TBMs depends to some extent on the logistics management of the consumables. The understanding of the importance of providing the consumables in the right amount at the right place and at the right moment is a key factor in this success. Logistics of consumables refers to the overall process of managing how the consumables are acquired, stored, and transported to their final destination. The shift TBM engineer is responsible for reporting to the site manager about the need for consumables. In simple terms, the goal of logistics management is to have the right amount of resources or input at the right time, get it to the appropriate place in the proper condition, and deliver it to the project site. The site manager is responsible for meeting shifting engineers' demands through the planning, control, and implementation of the effective movement and storage of related consumables, from the origin to the destination, i.e., tunnel face. For this, it is strictly important to have proper performance prediction of daily advance rates of TBM and project progress according to the Gant chart.

In this chapter, metro tunnels driven to Istanbul airport are taken as an example of the consumables of the tunnels having 6.6 m of external diameter. The consumables are first explained and later the amounts of consumables per m or m³ of tunnel driven are given to give clear ideas to the readers on this aspect of tunneling. The metro line with some of the general information about the project is given in Figure 7.1.

7.2 THE CONSUMABLES IN A TBM

Some of the materials consumed in the Istanbul airport metro drives are given in Table 7.1. The amounts of these materials per meter of tunnel drives are also summarized in this table. These values are obviously of prime importance in a management plan for consumables if daily advance rates are predicted based on ground conditions and the other factors governing TBM drives. The materials consumed are explained thereafter within the chapter.

DOI: 10.1201/9781003358978-7

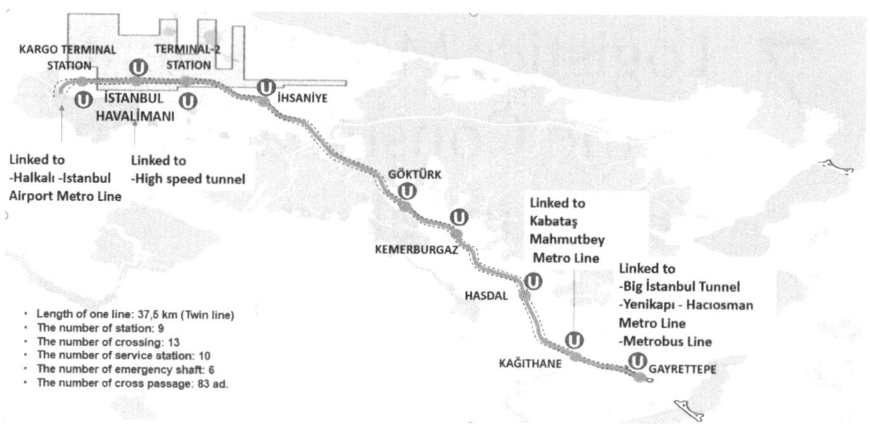

FIGURE 7.1   The outline of the Gayrettepe–Istanbul Airport Metro Line. (Cebeci 2021.)

7.2.1 Tail Greases and Steel Brushes

There is always a gap between the shield and precast segments as seen in Figure 7.2. Tail grease must effectively seal the gap between the shield and concrete segments to prevent the ingress of water, grout, or soil. It should resist high water and ground pressures and must have good pumping properties and adhesion to all surfaces. Protection of the tail brushes and tightness of the tail seal are essential in tunnel excavation. Tail greases are viscous, resistant to wash-out, and adhere to any damp surface. Pumped continuously as the TBM advances, sealing compounds for tail seals allow for good filling of the inter brush chambers to act as a barrier to water, grout, and soil fines coming from the ground. Before starting to bore, it is imperative that a first filling of the brushes should be carried out. The first-filling grease must be injected deep into the brushes until reflux on brush surfaces. The consistency and adherence of the first-filling grease ensure protection during the entire lifetime of the brushes. Sealing grease is the only material for a shield tail seal that is vital for shielded TBMs. In recent years, the construction of tunnels has consumed a significant amount of sealing grease, owing to the widespread use of TBMs. However, leakage often occurs at the shield tail, interfering with tunnel construction, and even causing accidents like in Foshan Metro Line 2 in China (Li et al. 2020; Yu et al. 2020). Generally, there are three or four rounds of steel brushes installed in the annular space filled by the sealing grease. The sealing grease is a necessary material to seal the gap between the shield shell and the segment and to prevent cement slurry or groundwater from entering the working space through the annular space. If the seal fails, the pressure of the backfill grouting will be insufficient, causing settlements on the ground surface and potentially even a collapse (Li et al. 2020).

TABLE 7.1
Material Consumed per Meter of Tunnel in Istanbul Metro Tunnel Drives

Line, Length, and TBM Used	Foam (Lt/m)	Tail Grease (kg/m)	HBW Grease (kg/m)	EP2 Grease (kg/m)
Line Kemerburgaz–Hasdal	74.16	36.81	2.45	2.70
Line1 5,413 m TBM NH-1	68.63	39.91	2.00	2.28
Line2 5,198 m TBM NH-2				
Line Kemerburgaz-E02	61.54	43.56	6.32	1.44
Line1 5,159 m TBM Lovsuns-1	52.99	38.54	–	2.21
Line2 5,142 m TBM Terratec				
Line Hastal–Kağıthane	59.00	51.18	9.45	2.45
Line1 5,251 m TBM Lovsuns-2	52.07	38.12	9.51	2.23
Line2 5,251 m TBM Lovsuns-3				
Line Kapıthane–Gayrettepe	65.63	46.63	–	1.64
Line1 3.008 m TBM	61.31	53.96		2.14
Herrenknecht S1084				
Line2 3.058 m TBM				
Herrenknecht S1082				
Line İhsaniye-T11	43.56	34.41	–	2.85
Line1 9,648 m TBM	46.23	28.82	–	2.87
Herrenknecht S1081				
Line2 9,612 m TBM				
Herrenknecht S1083				
Line İhsaniye-E02	49.65	28.79	–	3.16
Line1 6,859 m TBM	49.74	29.77	–	3.01
Herrenknecht S1084				
Line2 6,861 m TBM				
Herrenknecht S1082				

Source: From the archive of the author, S. Acun.

Note: EPB-TBMs from five different manufacturers are used in this project. The characteristics of these TBMs are given elsewhere within the book.

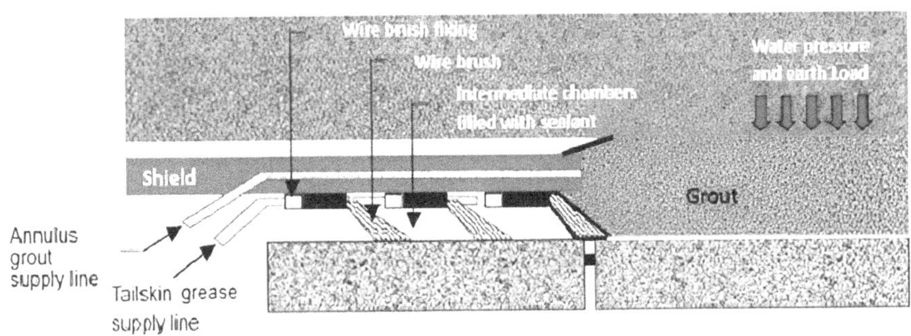

FIGURE 7.2 Schematic view of the use of tail greases, steel brushes, and grout between segments and TBM shield. (From the archive of the author, N. Bilgin.)

7.2.2 Main-Bearing Greases

Main-bearing greases protect the main bearing and prevent water, dust, and soil income from the main-bearing sealing. It must resist high water and ground pressures and must have good lubrication and pumping properties as well as excellent adhesion to all surfaces. The greases are available in several grades, with various thickeners and additives to reinforce the lubricant characteristics.

7.2.3 Fire-Resistant Fluids for Tunnel Boring Machines

In the underground environment where fire can have serious consequences for the personnel and equipment, the legislation in many countries requires fire-resistant hydraulic fluids.

7.2.4 Foaming Agents for EPB TBMs

Ground treatment is essential for the productivity and safety of the project. It is possible to transform the ground to get effective pressure control by injecting foam on the tunnel face. The treated ground shows lower permeability than untreated ground. These conditions allow excavation in complete safety while reducing torque, thrust force of TBM, and screw conveyor. They ensure better face pressure control, acting as an anti-clogging agent in sticky clayey formations, and reduction also in the wear of tools and their replacement frequency.

7.2.5 Foaming Agents for Rock TBMs

They act as anti-wear and anti-dust foaming agents for the ground treatment of rock TBMs. Rock TBMs create dust because of a much higher rotation speed of the cutting head and much drier spoil than earth pressure TBMs. Dust accumulation on the flanks of the disc cutters is a source of wear for the cutterhead, disc cutters, grippers, cutters, etc. Further, dust makes the atmosphere around the TBM harmful to the health of the workers.

7.2.6 Ground Treatment Additives and Polymers

Ground treatment additives and polymers are used in complex geologies such as the ground affected by fault zones, also in sticky, abrasive, or very permeable and supersaturated ground. The use of additives and polymers can provide effective treatment as a complement to the foam. These additives for foaming agents improve the productivity of TBMs. In the clayey ground, clogging of the cutters reduces the penetration rate of TBM creating higher torque and thrust values. In parallel with the foam, it is possible to use an anti-clogging agent which is diluted online in the foaming solution. In granular ground saturated in water, segregation between water and mud can occur, in this case, a polymer can be injected into the excavation chamber to correct the free water content and obtain homogeneous muck.

7.2.7 BENTONITES

During excavation, the use of bentonite mud aids the underground excavation. The slurry is prepared with the aid of bentonite mud which must meet certain specifications, according to the geology, field point, mud viscosity, etc.

7.2.8 GROUT AND GROUTING ADDITIVES

In a TBM drive, a gap that remains behind the shield tail between the lining segments and the surrounding ground is called the annular gap. This gap must be filled with a suitable material in order to ensure a uniformly distributed transfer of the loading from the ground pressure and also to counter any loosening of the surrounding ground. Annulus grouting is a key element in the safety of tunnel excavation. It limits the settlement by filling the annular gap left between the cutting diameter and the arching surface of the segments. It limits cave-in by filling the annular gap left between the cutting diameter and the external diameter of the segments. Grouting additives ensure the ability of long distance, pumping, and greater stability which limits the bleeding delayed setting and blocking of cement hydration with low environmental impact (Zhang et al. 2021). The mix design will determine the exact properties and behaviors of the grout to be used. Typically, single-component grout consists of cement, fly ash, sand, and bentonite. Grout material is mixed and transported into the tunnel, then pressurized with an injection pump and injected into the annulus. A two-component grout consists of an A component and B component. The A component is typically cement, sand, fly ash, and bentonite, while the B component is an accelerator derived from sodium silicate. In two-component grouting, the A and B components are kept in separate lines and mixed at the injection point into the annular gap (ITA report 2014). Component A is a cement mortar that is designed to be chemically and physically stable and characterized by long workability, commonly up to 72 h. Component B is an accelerator admixture that is added to the flow of component A just a few centimeters before the grout nozzle. It is typically a sodium silicate solution. Effectively, two-component grout technology, when correctly applied, allows a homogeneous, uniform, and immediate contact between the ground and lining (avoiding punctual loads on linings) and locks the rings into the designed position, avoiding movements due to both the segment's self-weight and thrust forces generated by the TBM advancement (Todaro et al. 2021).

Nowadays, two-component grout is the main technology used for backfilling and the uniaxial compressive strength is the key mechanical parameter used due to two aspects that are strictly linked. First, it is the key parameter imposed by designers in tunneling projects to characterize two-component grout at both short (some hours after the injection) and long (commonly 28 days after the injection) curing times. However, the creation of a gel is realized after a few seconds from the injection, with compressive strengths reaching from approx. 0.1 to 1 MPa at early stages. The gel exhibits a thixotropic consistency and starts developing mechanical strength almost instantaneously (weak but sufficient for the purpose: 50 kPa at 1 h is typical [Todaro et al. 2021]). Typical examples from Turkey of the mix of two-component grout

TABLE 7.2
The Mix of Two-Component Grout Used in Eşme–Salihli High-Speed Tunnel

Material	Volumetric (%)	
	Eşme High-Speed Tunnel	Istanbul Metro Airport Tunnels
Cement	8.04	10.3
Fly ash	11.3	8.1
Bentonite	0.44	0.4
Sand (0–3 mm)	53.2	63.2
Water	26.43	17.7
Plastizer	0.46	0.2

Sources: Kansu (2021) and data of Istanbul Metro Airport tunnels are provided by S. Acun.

Note: For Eşme Tunnel, a TBM having diameter of 13.7 m, rings of 1.8 m in length, and 0.45 m of thickness were used. Grout consumption for one ring was 14.2 m^3, or 7.9 m^3 per 1 m of tunnel. For Istanbul metro tunnels, the diameter of EPB-TBMs was 6.6 m and the mean grout consumption of 3 m^3 per 1 m of tunnel.

used in Eşme–Salihli High-Speed Tunnel, after Kansu (2021), and in Istanbul Metro Airport tunnels are given in Table 7.2

As explained by Pelizza et al. (2010) and Peila et al. (2011), *the two-component cementitious mix is typically a super fluid grout, stabilized in order to guarantee its workability for a long time (from batching to transport and injection), to which an accelerator admixture is added at the injection point into the "annulus". The mix gels a few seconds after the addition of the accelerator (normally 10–12 seconds, during which the TBM advances approximately 10–15 mm). The gel exhibits a thixotropic consistency and starts developing mechanical strength almost instantaneously (weak but sufficient for the purpose: 50 kPa at 1 h is typical). This system is injected under pressure throughout the "annulus" and is able to penetrate into any voids present. In addition, it can penetrate into the surrounding ground (depending on its permeability).* Water, cement, bentonite, and chemical admixtures are necessary to modify the water/binder ratio and the initial and final setting times constitute the cementitious mix. It is an active mix with very high fluidity. It has to be easily pumpable and is usually retarded (some hours) to avoid the risks of choking the pipes during transportation and injection (Pelizza et al. 2010; Peila et al. 2011). Two-component cementitious mix used per m^3 of hardened material in different projects in EPB-TBM drives is summarized in Table 7.3.

In the case of shielded hard rock tunnel construction with TBMs, the annular gap is filled with pea gravel. The grain size distributions between 4 and 16 mm are generally applicable. The pea gravel is used to immediately stabilize the segments. The pea gravel is usually screened alluvial pebbles, which are preferable, or crushed rock. The pea gravel is pneumatically pumped into the void through holes in the segments (normally the same holes used for the segment lifting device) or by blow pipes inserted

TABLE 7.3
Two-Component Cementitious Mix Used per m³ of Hardened Material in Different Projects in EPB-TBM Drives

Material	Metro Brescia Line 1 Italy	Metro Sofia Bulgaria	Auckland Sewer T. New Zealand	Metro Line C, Rome Italy
Water (kg)	816	795	730	770–820
Bentonite (kg)	42	45	30	30–60
Cement (kg)	315	290	480	310–350
Retarding agent (L)	3	2.5	1	3–7
Super plasticizer	–	–	5	–
Accelerator agent (L)	60	74	50	50–100

Sources: Peliza et al. (2010) and Peila et al. (2011).

between the tail shield and crown segment. The void with the injected pea gravel is then normally cement grouted as a secondary operation a few segments back from the tail shield. In order to avoid blockages within the annular gap, the percentage of the undersized grains should be kept below 10%. Also, the gravel should have no fines in order to minimize clogging. It is required that the "annulus" must be uniformly filled completely. Generally, it is considered that the larger the void content in the pea gravel accumulation, the larger the filling rate of the slurry in the backfill grouting, which means a better grouting quality and better workability of the mix. From the material point of view, the pea gravel material is a kind of a coarse aggregate with specific particle size, while the mixture of pea gravel and slurry is a special kind of concrete (ITAtech Report n°4 2014; Zhang 2021).

7.2.9   Wear Parts and Cutting Tools, Buckets, Scrapers, and Brushes

Disc cutters, buckets, and scrapers are the most essential parts of mechanical excavation, which control the efficiency and the cost of tunneling. It is strictly necessary to predict the amounts of these materials prior to starting the project for the control of logistics management. Some examples of the amount of these consumables in one of the most prestigious projects in Turkey are summarized in Table 7.4. For the other examples in other projects, the readers are advised to go through references from Bilgin et al. (2016).

7.3   CONSUMABLES IN A METRO PROJECT AND SOME IMPORTANT POINTS ON LOGISTIC MANAGEMENT

Some important information on Istanbul airport Metro Project and different consumables rather than described for TBM are given in Table 7.5. This table is important in the way that it points out basic information on logistics management of the consumables playing an important role in the time scheduling of the project.

TABLE 7.4
Number of Brushes, Disc Cutters, Buckets, and Scrapers per Meter of Tunnel Drives

Line, Length, and TBM used	Number of Brushes (per m)	Number of Discs (per m)	Number of Bucket/Scraper (per m)
Line Kemerburgaz–Hasdal	0.0000	0.0716	0.030
Line1 5,413 m TBM NH-1	0.0786	0.0658	0.009
Line2 5,198 m TBM NH-2			
Line Kemerburgaz-E02	0.1844	0.0464	0.000
Line1 5,159 m TBM Lovuns-1	0.1005	0.0402	0.004
Line2 5,142 m TBM Terratec			
Line Hastal–Kağıthane	0.1742	0.0903	0.016
Line1 5,251 m TBM Lovuns-2	0.2303	0.0843	0.033
Line2 5,251 m TBM Lovuns-3			
Line Kapıthane–Gayrettepe	0.2117	0.0568	0.025
Line1 3.008 m TBM Herrenknecht S1084	0.2004	0.0596	0.018
Line2 3.058 m TBM Herrenknecht S1082			
Line İhsaniye-T11	0.2486	0.0211	0.011
Line1 9,648 m TBM Herrenknecht S1081	0.2276	0.0128	0.013
Line2 9,612 m TBM Herrenknecht S1083			
Line İhsaniye-E02	0.2359	0.0215	0.024
Line1 6,859 m TBM Herrenknecht S1084	0.2090	0.0175	0.026
Line2 6,861 m TBM Herrenknecht S1082			

Source: From the archive of the author, S. Acun.

The success of a tunneling project lies in the proper project scheduling and the estimation of variation of job completion per year as seen in Figure 7.3. For this, it is strictly necessary to estimate daily advance rates, to formulate a Gant chart showing which part of the project is planned to be terminated at a given time. The accumulated data on consumables will definitely serve as a rational way to plan logistics management.

7.4 CONCLUSIVE REMARKS

As conclusive remarks, we share the thoughts of Scheffer et al. (2014) pointing out that projects in mechanized tunneling frequently do not reach their targeted production performance. Reasons are often related to undersized or disturbed supply-chain

TABLE 7.5
Some of the Different Items Used in Gayrettepe–Istanbul Metro Project with the Number of Each Item

The total length of metro lines	76,650 m	The deepest point from the ground	−71,97 m
The total length excavated by TBM	68,282 m	The steel used for strut and supporting elements	16,942 t
The length of tunnels excavated by NATM	8,309 m	Bored piles (6909)	188,451 m
The covered area of the nine stations	154,448 m²	The number of walking stairs	122
The excavated volume	5,295,469 m³	The number of elevators	45
The volume of the concrete used	1,557,822 m³	Fans for the ventilation	36
The amount of reinforcement and steel used	211,459 t	E&M transformers	
The amount of rail (158 km)	8,500 t	E&M panels	1.130
The number of machinery	549	The number of design sheets	41,924
The number of TBMs working at the same time	10		

Source: Cebeci (2021).

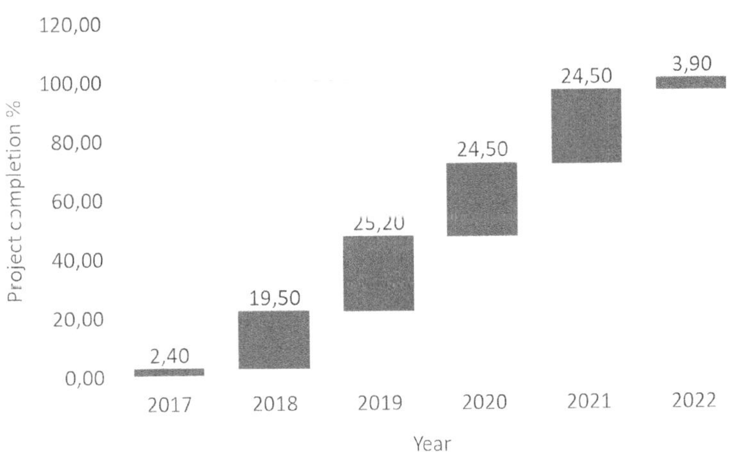

FIGURE 7.3 The variation of the percentage of the job completion time with year in Gayrettepe–Istanbul Airport Metro Project. (Cebeci 2021.)

management of the surface job site. Due to the sensitive interaction of production and logistics processes, planning and analyzing the supply chain is a challenging task. The understanding of the importance of providing the consumables in the right amount at the right place and at the right moment is a key factor in this success. Logistics of consumables refers to the overall process of managing how the consumables are acquired, stored, and transported to their final destination. In light of this, the amount per m of tunnel drive is given for foam, tail grease, main-bearing greases, steal brushes, discs, scrapers, and buckets for Istanbul metro drives, which is one of the biggest metro projects. The task of these consumables is explained within the chapter. The reader may extrapolate these results further for the other projects in order to have an estimate of the importance of the logistics management of the consumables in a TBM tunnel drive.

REFERENCES

Bilgin, N., Copur, H., Balci, C., 2016. *TBM Excavation in Difficult Ground Conditions*, Ernst & Sohn, Berlin, Germany, p. 329.

Cebeci, M., 2021. Istanbul-Airport metro project, presented by *International Tunnelling Symposium, organized by Turkish Tunnelling Society in Istanbul*. https://tunnelder.org.tr

ITAtech Report n°4 ITAtech, 2014. *Guidelines on Best Practices for Segment Backfilling*, International Tunnelling Assocition, p. 9.

Kansu, O., 2021. Ankara İzmir high-speed tunnel project, Eşme-Salihli T-01 tunnel, presentation submitted in *International Tunnelling Symposium held in Istanbul, organized by Turkish Tunnelling Society*. https://tunnelder.org.tr

Li, X., Yang, Y., Li, F., 2020. Comparative study on mechanical properties of sealing grease composed of different base oils for shield tunnel. *Materials* 13(692), pp. 1–13, doi:10.3390/ma13030692

Peila, D., Borio L., Pelizza, S., 2011. *The Behavior of a Two-Component Backfilling Grout Used in a Tunnel-Boring Machine*, Acta Geotechnica Slovenica, University of Maribor, Faculty of Civil Engineering, 1, pp. 5–15.

Pelizza, S., Peila D., Borio L., Dal Negro, E., Schulkins, R., Boscaro, A., 2010. Analysis of the performance of two-component back-filling grout in tunnel boring machines operating under face pressure, *ITA-AITES World Tunnel Congress 2010 Vancouver*, 14 to 20 May, Canada, p. 8.

Scheffer, M., Rahm, T., Duhme, R., Thewes, M., König, M., 2014. Job logistics in mechanized tunneling, in *Proceedings of the 2014 Winter Simulation Conference*, A. Tolk, S.Y. Diallo, I.O. Ryzhov, L. Yilmaz, S. Buckley, and J.A. Miller (eds), IEEE, Savannah, GA, USA, pp. 1843–1854.

Todaro., C., Andrea Carigi, A., Martinelli, D., Peila, D., 2021. Case study: Study of the shear strength evolution over time of two-component backfilling grout in shield tunneling. *Case Studies in Construction Materials,* 15(00689), pp. 1–14.

Yu, C., Zhou, A., Chen, J., Arulrajah, A., Horpibulsuk, S., 2020. Analysis of a tunnel failure caused by leakage of the shield tail seal system. *Underground Space*, 5, pp. 105–114.

Zhang, J.L., Huang, Q.X., Hu, Ch., Wang, Zh.Q., 2021, *Experimental Investigation into Material Characteristics of Pea Gravel, Archives of Civil Engineering*, Warshaw University of Technology, pp. 15–435, doi:10.24425/ace.2021.138063

8 Job Organization in Mechanized Tunneling with a TBM

8.1 INTRODUCTION

A good organization of the project management, efficient utilization of the labor, material, and equipment are key factors in the success of a tunneling project. Since labor is a large part of the tunneling cost, the interaction between management organization and the number of workers associated with this will play a major role in the overall success of the project. Taking into consideration this important issue, this chapter is mainly focused on the description, qualifications, skills, and the role of the key staff in the organization chart and the number of workers employed in a well-determined metro project. As an example, Gayrettepe–Istanbul Metro Project will be taken as a basis to describe basic concepts in the management of job organization. This project is well described in some of the chapters of this book. However, some key points are as follows: the length of the metro line is 37.5×2 (double line) km, the number of stations is 9, the number of crossings is 9, the number of service stations is 10, the number of emergency shafts is 6, and the number of crosscuts is 83. The project was started in the last months of 2017 and finished at the beginning of 2022. Ten EPB-TBMs were used in the project. About 68.282 m of tunnels were excavated with TBMs and 8,309 m of tunnels with conventional tunneling methods (NATM). 5.295.469 m³ of material was excavated throughout the project for opening the tunnels, stations and for enviromental lanscaping..

8.2 DESCRIPTION OF THE KEY STAFF IN A MECHANIZED TUNNELING PROJECT

The qualifications, roles, and skills of the project staff determined in the above-mentioned project are described in the following.

8.2.1 PROJECT MANAGER

The executive committee assigned by the joint venture appoints the project manager who is responsible for the management and supervision of the works being carried out and defines the technical and management aspects of the project. He reports directly to the executive committee. The following abilities and qualifications are demanded

DOI: 10.1201/9781003358978-8

from a project manager: an engineering degree preferably in construction with an M.Sc. degree. It is also preferable that he has followed basic courses in business administration and economics. He should have ten years of experience in similar projects, leadership and management skills, the ability to represent the joint venture, communication management skills, analytical thinking, analysis and synthesis skills, and team-building skills. He is also required to have skills in quality management, process management, scheduling/time and resource management, risk management, financial reporting, general knowledge of contract law, and experience.

8.2.2 Project Deputy Manager

The project deputy manager is appointed by a joint venture and reports to the project manager. He is asked to have five years of experience in the same type of job and to have the same qualifications and talents as the project manager.

8.2.3 Quality Control Manager

He is appointed by the project manager and he reports directly to him. He should have an engineering or Bachelor of Science degree with five years of experience in management systems. He should have certificates of ISO 9001, ISO 14001, and OHSAS 18001 and set up, realize, and control management systems on quality control, environmental control, and worker health and safety.

8.2.4 The Worker Health and Safety and Environment Manager

He is appointed by the project manager who reports directly to him. He must have the same qualifications and talents that the quality control manager has. He is responsible for evaluating the risks connected with the various activities and preparing a safety plan, which he updates continuously in relation to the varied site needs. He is responsible for the training and information of the workers and for all the other procedures required by the law. He also collects and keeps safe all the certifications, all the user manuals, and wherever necessary for the proper management of the plant and equipment present on site.

8.2.5 Technical Office Manager

He is appointed by the project manager and he reports directly to him. He should have five years of experience in similar projects. He should preferably have a civil engineering degree with leadership and management skills, analytical thinking and analysis/synthesis ability, team-building skills, planning (time and resource management) skills, general contract law knowledge, and experience.

8.2.6 The Manager of Financial and Administrative Section

He is appointed by the executive committee and reports directly to the project director. He should preferably have a business or economics degree and ten years of experience in his field. He is asked to have the ability of analytical thinking and problem-solving ability.

8.2.7   THE MANAGER OF ELECTRO-MECHANICS SECTION

He is appointed by the executive committee or by the project director and he is asked to have ten years of experience in his field and experience. He reports to the deputy project manager.

8.2.8   THE MANAGER OF PROJECT PLANNING AND DESIGN

He is appointed by the project manager or deputy manager, and he reports directly to the project deputy manager. He should preferably have a civil engineering degree with at least five years of experience in structural design and modelling, knowledge of foreign languages, leadership, management, coordination skills, analytical thinking and analysis, synthesis ability, team-building skills, planning (time and resource management) skills, knowledge and experience in general contract law, and experience in FIDIC applications.

8.2.9   THE TUNNEL MANAGER

He is appointed by the project manager or deputy manager and reports directly to the project deputy manager and is responsible for the planning, budget, all performance of all the underground works, all the preparatory work, and the supplies necessary to work.

8.2.10   THE TBM GROUP MANAGER

He reports directly to the tunnel manager and is in charge of the appropriate management of all personnel and plant and equipment employed for the tunnel excavation. The TBM superintendent supervises both production and plant maintenance, plans with the workshop master mechanic the maintenance under his supervision, and guarantees during shift change that the information is correctly transferred between corresponding personnel.

8.2.11   THE SHIFTS OF SUPERINTENDENTS/FOREMEN

They are responsible for carrying out the following operations, under the supervision of the machine superintendent: organization of the work activities strictly connected with the excavation, precast lining segments, ring erection, and grout backfill behind the lining.

8.2.12   THE MANAGER OF SURVEYING AND GEOTECHNICAL SECTION

He is appointed by the project deputy director who reports directly to him. He should have at least five years of experience: a degree in geology, geodesy, and mining, or civil engineering.

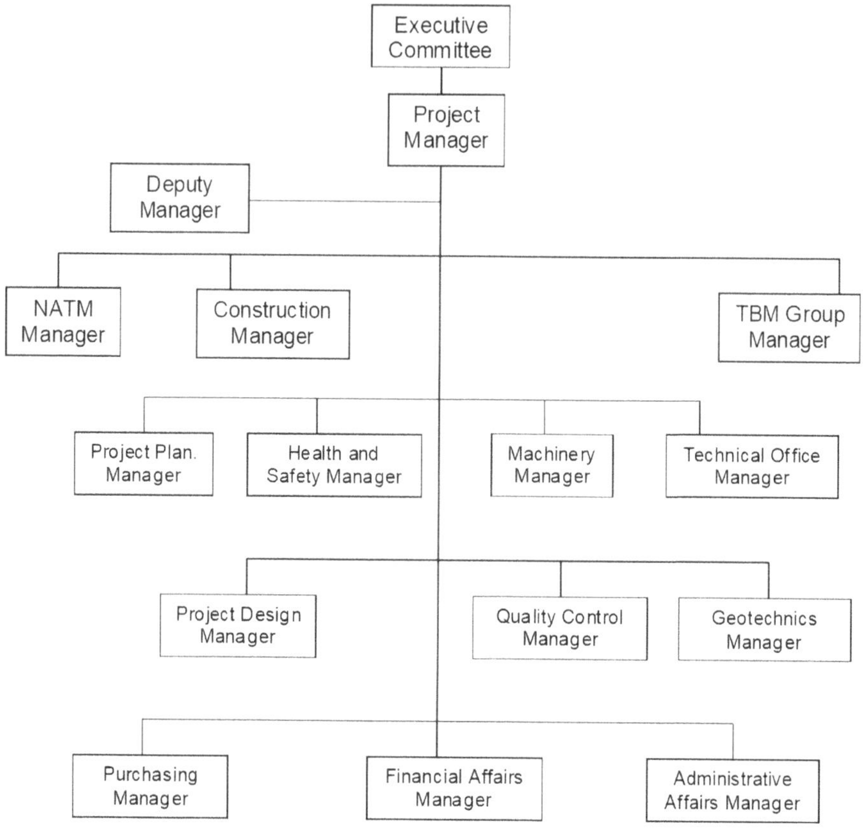

FIGURE 8.1 The flow chart of the organization for the project management in Gayrettepe–Istanbul Metro Project.

The flowchart of the organization for the project management of Gayrettepe–Istanbul Metro Project is given in Figure 8.1.

8.3 JOB AND CREW ORGANIZATION

Job and crew organization is one of the most important factors in the efficient execution of a project, especially in densely labor-dependent projects like tunneling. As an example, at peak working time (in November 2020), a total number of 4,762 people were employed in the Gayrettepe–Istanbul Airport Metro Project, of which 2,100 people were employed as contractors and 2,662 people were employed as subcontractors. The mean numbers of workers on the different campuses of Gayrettepe–Istanbul Airport Metro Projects are given in Table 8.1. The distribution of employees/workers and the distribution of employees/workers within each working group are given in Figures 8.2

TABLE 8.1
Mean Numbers of Workers in the Different Campuses of Gayrettepe–Istanbul Airport Metro Project

Campus Area	Covering Area	Number of Workers
Gayrettepe Northern Campus	1.057 m²	256
Gayrettepe Central Campus	1.206 m²	216
Hasdal Jobsite Campus	11.680 m²	732
Hasdal Central Campus	39.648 m²	97
Kemerburgaz Campus	10.083 m²	982
Boğaziçi Segment Campus	2.606 m²	110
Göktürk Campus	2.409 m²	96
İhsaniye Campus	18.041 m²	1.000
Central Campus	8.104 m²	608
Total	**94.834 m²**	**4.097**

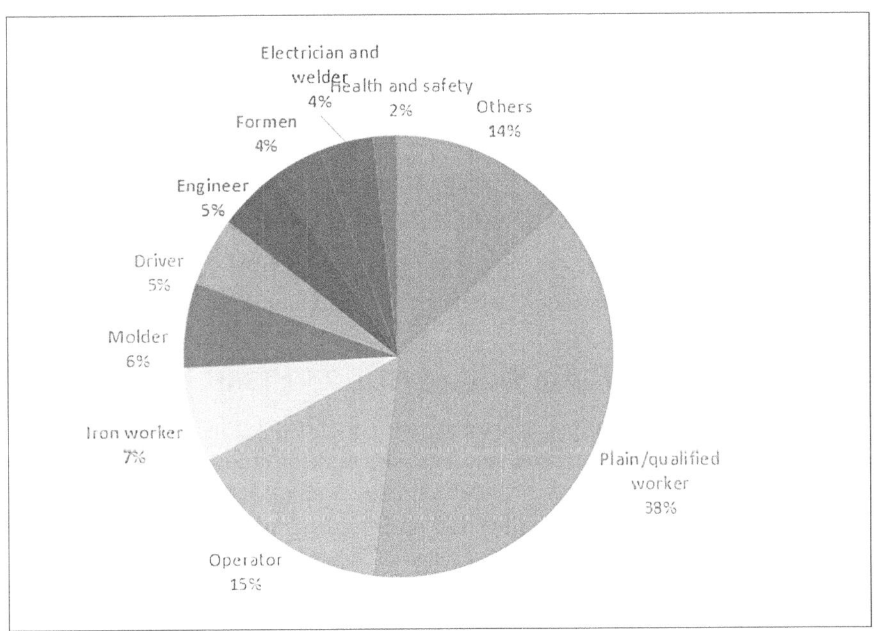

FIGURE 8.2 Distribution of employees/workers in Gayrettepe–Istanbul Airport Metro Project.

and 8.3. The most important point emerging from these figures is that 38% of the working group are simple or qualified workers and 15% are operators. Forty-three percent of the people work in NATM (conventional tunnel driving in shafts, stations, etc.), 28% in TBM drives, and 11% are white-collar workers (Cebeci, 2021)

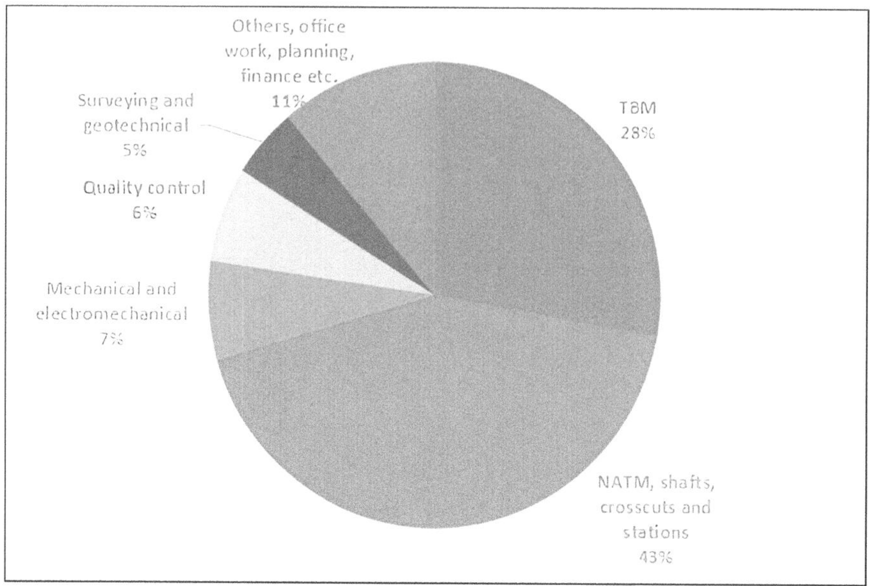

FIGURE 8.3   Distribution of employees/workers within each working group in Gayrettepe–Istanbul Airport Metro Project.

8.4   THE ORGANIZATION OF TBM CREW, DEPENDENCE ON TBM DIAMETER, TYPE, AND GROUND CHARACTERISTICS

As practice shows, the crew size in a TBM drive depends on TBM type hard rock/EPB/slurry, on ground conditions, and partly on TBM size.

8.4.1   TBM Crew in a Small Diameter Hard Rock TBM

In the 13.5 km Amlach tunnel in Austria, the crew of a 3.9 m diameter hard rock gripper TBM consisted of 14 men as given in the following (Janzon and Buechi 1987):

1 shift foremen, 1 TBM operator, 1 electrician, 4 miners/laborers, 1 muck loader, 2 loco drivers, 1 switchman, also a fan and water pipes, 2 laborers at the portal, on/off loading, 1 mechanic. The workshop consisted of 7 men on the day shift and there was also a staff of 7 persons on site. In this project, the mean utilization time was 49% of the total working time. The average daily advance rate was in excess of 40 m with the best day of 83 m in dolomitic limestone.

8.4.2   The Organization of the TBM Crew in a Medium Size Rock TBM

The effect of hard rock TBM diameter on crew size is reflected in the reports given by Neil and Dalton (1981). The number of hard rock TBMs used in the Metropolitan Sanitary District of Greater Chicago was 11 with diameters changing between 4.65– and 10.77 m. TBMs worked mainly in dolomites with chert and shale reaching mainly an advance rate of 18.4 m +/– 3 s.d. The average crew size was 20 men +/– 4.

8.4.3 The Organization of the TBM Crew in a Single Shield TBM of 11.2 m Diameter

The crew size given by Chavan et al. (1989) for a single shield TBM of 11.52 m diameter working in the Zurichebeerg Tunnel in sandstone, marl, and siltstone was in the order reported by Neil and Dalton (1981). A mean daily advance rate of 11.9 m was reached and the crew size was 23 assigned working elements per shift as given in the following:

1 tunnel walker/foremen, 1 TBM operator, 1 conveyor operator, 3 muck track operator, 1 muck dumpsite dozer operator, 1 track relying crane operator, 1 erector operator, 2 place precast segment, 1 bring segment from the rear, 2 precast segment delivery truck, 2 grout and gravel pumping, 1 backfill foreman, 1 loader operator, 1 compactor operator, 2 backfill dump truck operator, 2 subdrain placement.

8.4.4 The Organization of the TBM Crew in Double Shield TBM of 11.7 m Diameter Driving in Difficult Conditions

Hsuehshan Tunnel driven in Taiwan was among the most difficult TBM projects at its time in the world in terms of adverse geology, water inrushes, tunnel size, and tunnel length, which is composed of two main tunnels (westbound and eastbound) and a pilot tunnel. A double shield of 4.8 m diameter was used for the pilot tunnel 5 m beneath the Eastern and Western tunnels and two identical double shield TBM of 11.7 m diameter were used in the main tunnels (Lee and Leng 2005). Therefore, we summarized in Table 8.2, working crews of the TBM construction team (engineers and labor) for one shift in the Eastern tunnel. We think that it is a good example to some difficult projects.

8.4.5 Crew Organization of a 6.7 Diameter EPB-TBM Drive

The following numbers are for one shift in Gayrettepe–Istanbul Airport Metro Project:

2 engineers: civil, mining, mechanical, geologist, or geophysicist, 2 foremen, 2 pipefitter, 2 welder, 2 electricians, 7 operators: TBM, segment erector, MSV, etc., 20 labors.

8.4.6 Crew Organization of a 13.77 m Diameter EPB-TBM Drive in Eşme Tunnel/Turkey

In Eşme Tunnel, a total of 72 workers were working within the tunnel in two shifts, 36 being in one shift. The distribution of workers in one shift is given in the following:

2 engineers: civil, mining, mechanical, geologist, or geophysicist, 2 foremen, 3 pipefitter, 1 welder, 4 mechanic, 4 electricians, 7 operators: TBM, segment erector, locomotive, etc., 13 laborers.

Forty-six people were working outside of the tunnel. However, in plus to this 120 people were employed in segment plant, 50 iron workers, 50 mold workers, and 30 people charged with maintenance, electricians, and operators of mixers and crane (Kansu 2021)

TABLE 8.2
Eastbound Main Tunnel TBM Construction Team

Technical People Working Under Mechanical Shift Engineer	Technical People Working Under Civil Shift Engineer	Technical People Working Under Electrical Shift Engineer
6 labors working as cutter inspecting and changing the cutters, hydraulic system maintenance and repair, conveyor maintenance and repair, miscellaneous works *2 labors* along the tunnel and outside portal area *6 labors* working as maintenance and repair of rails and tracks	*2 workers* working in segment erection and in muck cleaning *4 workers* working in pea gravel buck filling, mortar grouting, steel rib installation (if necessary) *2 workers* in miscellaneous works, 2 as crane operator *4 workers* in rail installation, etc. *11 workers* as locomotive drivers *2 workers* as coordinator and miscellaneous works	*2 workers* as PLC and electronic system maintenance and repair *1 worker* as electrical system maintenance and repair *1 worker* as conveyor operator *2 workers* as dumping device operator and miscellaneous works.

Source: Lee et al. (2005).

Note: TBM was a 11.7 m diameter with hard rock TBM. At the last stage while TBM was working deep into the mountain for about 7 km, 55 labors were working per shift of 8 h. The above did not include working crew sitting outside portals such as rolling stock workshop, precast loading and unloading yard, cutter workshops, and so on.

8.5   CONCLUDING REMARKS

A contractor who is charged with a tunneling project should know that his success will mainly depend on the experience, qualifications, and skills of his employees. A good organization of the project management, efficient utilization of the labor, material, and equipment will be his indispensable and essential actions. This chapter is devoted to understanding the basic concepts of job and crew organization with TBM drives. Several examples are given on hard rock, single shield, double shield, and EPB–TBMs with different diameters, working in different ground conditions. The most important points emerging from these figures are, in a big-sized metro project, 38% of the working people group are simple or qualified workers and 15% are operators. Forty-three percent of the people work in NATM (conventional tunnel driving in shafts, stations, etc.), 28% of the people work in TBM drives, and 11% of the people are employed as white-collar workers. It is also reported that the size of the tunnel, the type of ground, and TBM play an important role in assigning the TBM crew.

REFERENCES

Cebeci, M., 2021. Istanbul-Airport metro project, presented made by Murat Cebeci, the project director, *International Tunnelling Symposium organized by Turkish Tunnelling Society in Istanbul*. https://tunnelder.com.tr

Chavan, F., Ritz, W., Windler, H.J., Yanagisawa, S., 1989. Zurichbeg railroad tunnel, *Rapid Excavation and Tunnelling Conference*, June 11–14, Los Angeles, California, pp. 663–677.

Janzon, H., Buechi, E, 1987. Record performance of TBM in the 13.5 km Amlach Tunnel, *Austria, Rapid Excavation and Tunnelling Conference*, June 14–18, New Orleans, Louisiana, pp. 1251–1268.

Kansu, O., 2021. Ankara İzmir high-speed tunnel project, Eşme-Salihli T-01 tunnel, presentation submitted in *International Tunnelling Symposium Held in Istanbul, Organized by Turkish Tunnelling Society*. https://tunnelder.com.tr

Lee, W.C., Leng., Y.T., 2005. Management of TBM construction in Hsuehshan main tunnels and pilot tunnel, *Proceedings of the International Conference on World Long Tunnels 2005*, November, Taipei, Taiwan, pp. 223–236.

Neil, C.F., Dalton, F.E., 1981. Building the tunnel and reservoir plan using the contracting practices of the metropolitan sanitary district of greater Chicago, *Rapid Excavation and Tunnelling Conference*, May 3–7, San Fransico, pp. 162–650.

9 Segment Production

9.1 INTRODUCTION

TBMs drives are mainly associated with precast concrete segment tunnel linings consisting of rings sequentially placed as the tunnels advance. Each tunnel project has special tunnel lining requirements, depending on the diameter, the ground conditions, and the other characteristics of the tunnel, such as curvature. Perfectly manufactured segments must be supplied just in time for efficient tunnel production. After their installation in the tunnel, they have to withstand demanding use for up to 100 years. The segments put on the rock wall are joined together in order to form a concrete ring. They have several roles within the tunnel, allowing TBM to move forward, as the machine leans upon the last ring to progress inside the tunnel, finishing up the internal wall of the tunnel and ensuring that the tunnel will resist ground pressure. The concrete ring is usually composed of a variable number of segments (from 4 to 10), depending on the tunnel geometry and constraints. Some elements of the segments, stocked up segments, and TBM thrust cylinders pushing the segments are seen in Figures 9.1, 9.2, 9.3, 9.4, and 9.5.

9.2 SEGMENT PRODUCTION

Segment production is a major part of a mechanized tunnel project and, as seen in one of the chapters of this book, the cost of the precast segments may reach half the cost of all the equipment used in the project. The TBM has to be constantly supplied with segments and for this, a precast plant is set up, either stationary or automated through a carousel. The choice between stationary and carousel is either economical or depends upon the project's progress. Therefore, several molds are needed to create a complete ring. All the molds needed to compose one ring from a set of molds. The mold plays an essential role in the creation of a tunnel using segments. The plant, located next to the tunnel project, has to be adapted in order to fit in the molds, as well as the handling equipment necessary for the completion of the project. In a carousel, the molds are moving while the workers and equipment are not (Jiang, 2019; Praneshwari, 2019). A drive system allows molds to go from one station to another with the help of rails on the working line. All the necessary steps to produce a segment are carried out on this line, such as mold preparation, concreting, demolding, and cleaning. After

DOI: 10.1201/9781003358978-9

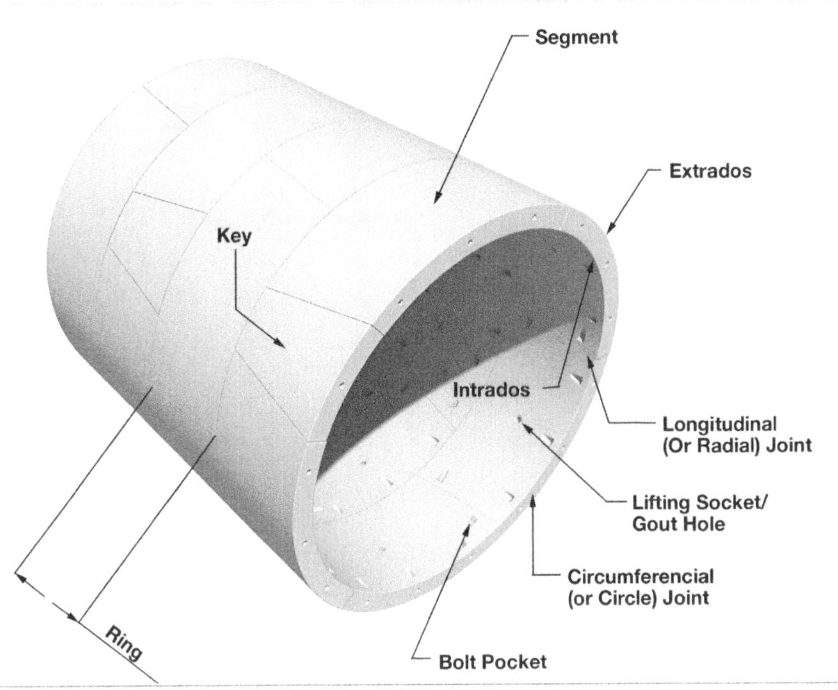

FIGURE 9.1 Some elements of a precast segment. (Hurt, 2019, https://tunnelingshortcourse. com/2016-presentations/hurt-segmental-concrete-liners.pdf)

concreting, the molds are placed in the curing room in order to dry faster where the maximal temperature is 65°C and the humidity is close to 80%. Once demolded with the help of a vacuum lifter, the precast elements are transferred to the evacuation line, in order to place additional devices such as gaskets, springs, and inserts. They are then stored in a dedicated area, waiting to be moved to the TBM. The production of a single segment, going through all these steps, usually takes 7 hours, of which 6 hours in curing chamber. In a carousel, a new segment is produced for around 10 minutes.

In a stationary plant, molds are fixed to the floor for the whole duration of the project, while workstations are mobile. All the equipment, concrete included, has to be dispatched to the molds. Operators also have to move to complete each step of the segment production process.

The main advantages of a carousel plant over a stationary plant are the increase in productivity and the need for fewer workers. However, Riechers (2014) comparing several projects, concluded that the decision on the selection between carousal and stationary plants depends on numerous infrastructural and economic aspects and should be based on a thorough evaluation of all criteria relevant to the corresponding project. A view of a stationary plant from Eşme Tunnel, Turkey and a view of carousel from Kemerburgaz work site in Gayrettepe Istanbul Airport Metro Project are seen in Figures 9.6, 9.7, 9.8, and 9.9.

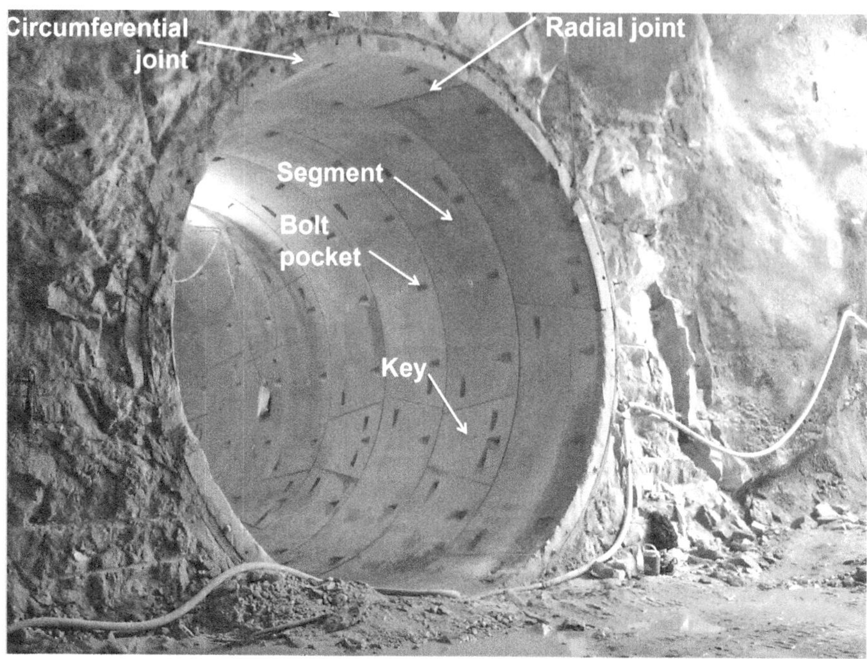

FIGURE 9.2 Some elements of a precast segment are seen within a tunnel. (Hurt, 2019, https://tunnelingshortcourse.com/2016-presentations/hurt-segmental-concrete-liners.pdf)

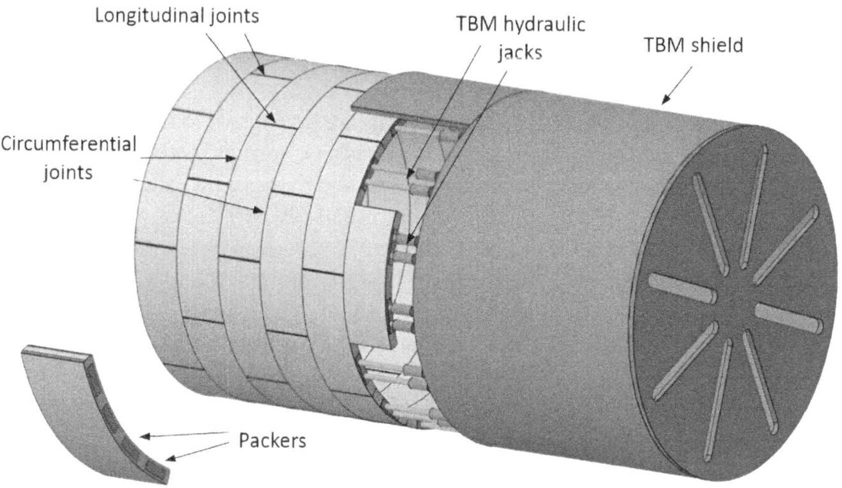

FIGURE 9.3 Schematic view of precast segments set up within TBM. (Hurt 2019, https://ori olarnau.wordpress.com/home/segmental-tunnel-linings/)

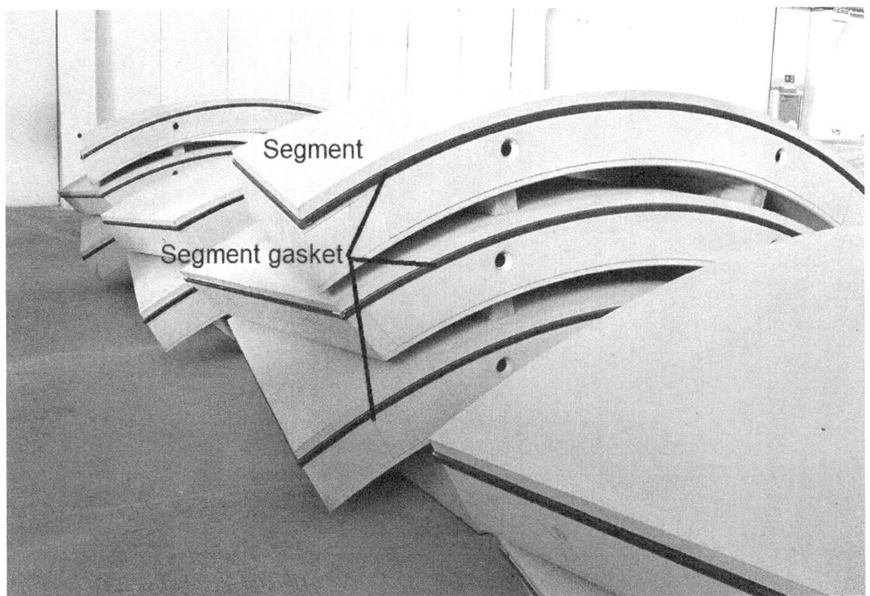

FIGURE 9.4 Segments stocked up in a proper way, in a tunnel yard. (Simons, 2020.)

Material consumption is one of the most important factors in tunneling cost. Consumables per ring in a metro project in India are given in Table 9.1 to give an idea to readers of this book.

9.3 MANPOWER AND THE SPACE NEEDED FOR SEGMENT PRODUCTION AND STORAGE

The basic tasks of the manpower in segment production are, cleaning and oiling the mold, producing reinforced steel cages, concreting, finishing, curing, stripping, controlling and repairing, placing the gaskets, and segment storage. Table 9.2 is a summary of manpower per shift, ring production per shift, number of molds, and indoor/outdoor areas in Kemerburgaz, jobsites of Gayrettepe Istanbul Airport Metro Project. As seen from this table, the ratio of manpower to ring production per shift is 2 men/ring in the carousel established in Kemeburgaz. However, this ratio is 4.75 men/ring for the segment production using a stationary plant, resulting in an efficiency of almost twice if manpower is considered. At the other end, this table shows that the ratio of indoor area to ring production per shift is 133.3 m^2/ring in a carousel segment plant. However, this ratio is 350 m^2/ring for the stationary segment plant. These findings clearly support the views given by Riechers (2014). For a comparative study, the design, production, and consumable parameters for bigger-size segment production in a carousel plant in Eşme high-speed railway tunnel are given in Tables 9.3 and 9.4.

The total weight of one ring is 22,156 kg (density of concrete being 2.5 kg/m^3), with 8.42 m^3 of concrete and 1,106 kg of reinforcement.

FIGURE 9.5   TBM thrust cylinders pushing the segments. (Simons, 2020.)

9.4   QUALITY MANAGEMENT PLAN IN SEGMENT PRODUCTION

The quality management (QM) plan provides guidance on how quality will be ensured in the project through design reviews, documentation, and other protocols. It gives management and the customer a clear understanding of how quality will be maintained and what documentation they can expect (addressing quality) during the life of the project.

FIGURE 9.6 A view of a stationary plant from Eşme Tunnel, Turkey. (From the archive of the author, S. Acun.)

FIGURE 9.7 A view of the carousel from Kemerburgaz work site in Gayrettepe Istanbul Airport Metro Project. (From Cebeci, 2021.)

For the production, storage, and transport of the segments, a project-based QM and testing plan should be created and used for each tunnel project. This QM plan contains the specifications of the client and the contracting company for the durable, fault-free, certified and tested production, and delivery of the segments. The construction operation requirements of the segments from delivery on the site until installation in the tunnel are defined in the QM plan of the shield manual (Daub, 2013).

FIGURE 9.8 A view of a stationary plant from Eşme Tunnel, Turkey. (From the archive of the author, S. Acun.)

FIGURE 9.9 A view of a carousel from Kemerburgaz work site in Gayrettepe Istanbul Airport Metro Project. (From Cebeci, 2021.)

TABLE 9.1
Consumables per Ring

Items Description	Unit	Consumption per Ring
Grout socket, 56 mm	Number	6
Black casting socket	Number	28
Concrete, M_{50}	m^3	7.5
Reinforcement, 8 mm	kg	192
Reinforcement,10 mm	kg	241.4
Reinforcement, 12 mm	kg	195.4
Binding wire	kg	7
Shuttering oil	L	2
Bubble sheet for curing	m^2	2.5
White cement for micro finishing	kg	3.8

Source: Singh (2018).

Note: 7.5 m^3 is consumed for 1 ring consisting of 5 segment + 1 key, the inner diameter is of 5.8 m, the thickness is of 27.5 cm, and length is of 1.4 m.

TABLE 9.2
Manpower per Shift, Ring Production per Shift, Number of Molds, and Indoor/Outdoor Areas

Segment Plant	Indoor/Outdoor Areas (m^2)	Manpower per Shift	Production Ring per Shift	Number of Molds
Carousel, Kemerburgaz	2,000/2,450	30	15	60 for 10 rings
Stationary, İhsaniye	7,000/5,500	95	20	168 for 28 rings

Source: From the archive of the author, N. Bilgin.

Notes: Kemerburgaz and İhsaniye are two jobsites of Gayrettepe Istanbul Airport Metro Project. Segment outside diameter is of 6.3 m, segment outside diameter is of 5.7 m, and segment length is of 1.5 m.

The QM plan for segment production should address the following points:

Production – Steps

Cleaning and oiling of the formwork, inserting the reinforcement, closing the formwork and mounting installation parts, final inspection of the formwork before concreting, concreting and compaction, opening and cleaning the counter formwork, stripping and smoothing the segment back, concrete curing (up to stripping), heat tunnel in carousel production, lifting off the segment.

TABLE 9.3
Crew in a Carousel Segment Plant in Eşme

Employee	Number
Manager of the plant	1
Shift engineer	1
Formen	1
Electrician	1
Operator of the crane	3
Welder	6
Iron worker	25
Concrete worker	7
Unskilled worker	7
Driving of the moving mixer	3
Driver of the forklift	1
Fitter	1
Labor charged of the gasket	3
Driver for segment transport	2

Source: From the archive of the author, S. Acun.

Note: The number of the crew is 61 per shift, the inside door area is of 4,500 m², the outdoor area is of 5,500 m², the number of rings per shift the number of molds, 4 sets.

TABLE 9.4
Design Parameters of a Ring in Eşme Tunnel with the Number of Consumables per Ring

Ring Design	$8 + 0$
The length of the ring	1.8 m
The outside diameter of the ring	13.4 m
The inside diameter of the ring	12.5
The thickness of the segment	450 mm
The amount of concrete per ring	32.8 m³
The amount of reinforcement per ring	3,280 kg

Source: From the archive of the author, S. Acun.

Curing

Fresh concrete temperature, the temperature of the production site, temperature profile.

Interim storage/maturing storage

Supports and interim storage battens, storage time and necessary concrete maturity, handling and transportation equipment, protection against drying out and drafts, tolerance control, joint control.

Concrete repair

Cracks, minor damage and air voids, damage to the segment outer/inner side to the reinforcement, damage to the segment inner side to the sealing joint with small/large dimensions, refurbishment concept and documentation.

Open air storage

Foundation and stability, supports and interim storage battens, space requirements, handling, and transportation equipment, the configuration of the segment stack, weather protection against snow, ice.

Materials for concrete

Materials such as aggregates, binders, additives, admixtures; synthetic fibers, concrete supply, and equipment.

Materials for reinforcement and installation parts

Steel bars, steel fibers, concrete cover and spacers, dowels, etc.

Materials for release agents, curing agents, and materials for concrete repair with different trade names

Test plan, factory acceptance and quality testing segment molds commissioning production plant, internal and external monitoring of the concrete, quality control of the reinforcement cages, incoming inspection and quality assurance of the suppliers, ongoing control of formworks, segment measuring, sample ring, documentation/traceability of segments.

Occupational safety

Noise, concrete contact, dust, handling hazardous materials, handling equipment.

9.5   QUALITY CONTROL

All segments are systematically checked for surface defects and repaired if required. In the production plant and in the TBM worksite facility the following control and acceptance procedure are usually realized. If a segment has broken edges of more than 25 mm in dimension, blowholes between 5 and 8 mm in diameter, cracks with a depth more than 20 mm, honeycombing having a depth of more than 20 mm, and broken edges of more than 25 mm in dimensions, it is rejected and is removed from

FIGURE 9.10   Honeycombing in the segment. (From the archive of the author, N. Bilgin.)

the worksite. Blowholes are individual, generally rounded, cavities in the concrete, generally less than 10 mm across. They are caused by air in the concrete being trapped against the form face, sometimes due to insufficient vibration (see Figure 9.10). Honeycombing refers to voids in concrete caused by the mortar not filling the spaces between the coarse aggregate particles (see Figure 9.11).

However, any defects occurring to the segments once they constitute part of the tunnel lining are repaired under the acceptance and the control of the chief tunnel engineer This case also includes the leaky cracks, as well as the leaky joints between the segments or the ring.

An appropriate and approved epoxy is used for cracks on erected segments. An acrylic or polyurethane resin is used on erected leaky segments, and polyurethane foam is used if there is a high-pressure water leak.

The segments are made of C45/C60 concrete. Slump test is one of the important tests for concrete. It is desirable that the segments to be cast with concrete have slump values of 7–9. For higher values of slump values, segments will have rough or uneven surface like seen in Figure 9.12.

Readers are advised to read the papers by Shayanfar et al. (2017) for the classification of precast concrete segments damages during production and transportation, Hariyanto et al. (2005) for the quality control in precast production, Abbas (2014) for structural and durability performance of precast segmental tunnel lining, and ITA (2019) Report No 22, for guidelines for the design of segment tunnel linings.

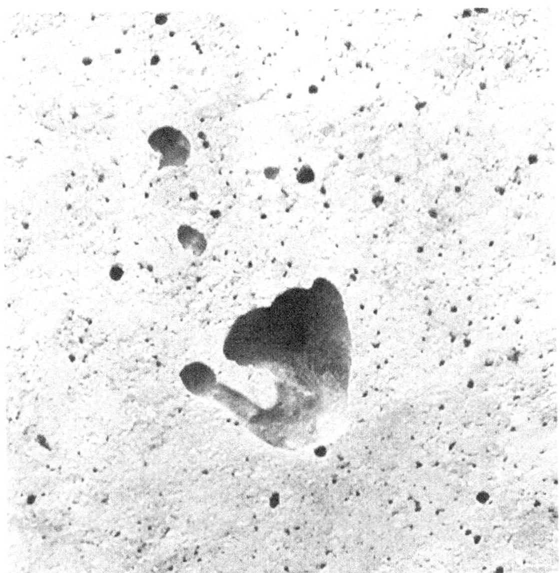

FIGURE 9.11 Honeycombing in the segment. (From the archive of the author, N. Bilgin.)

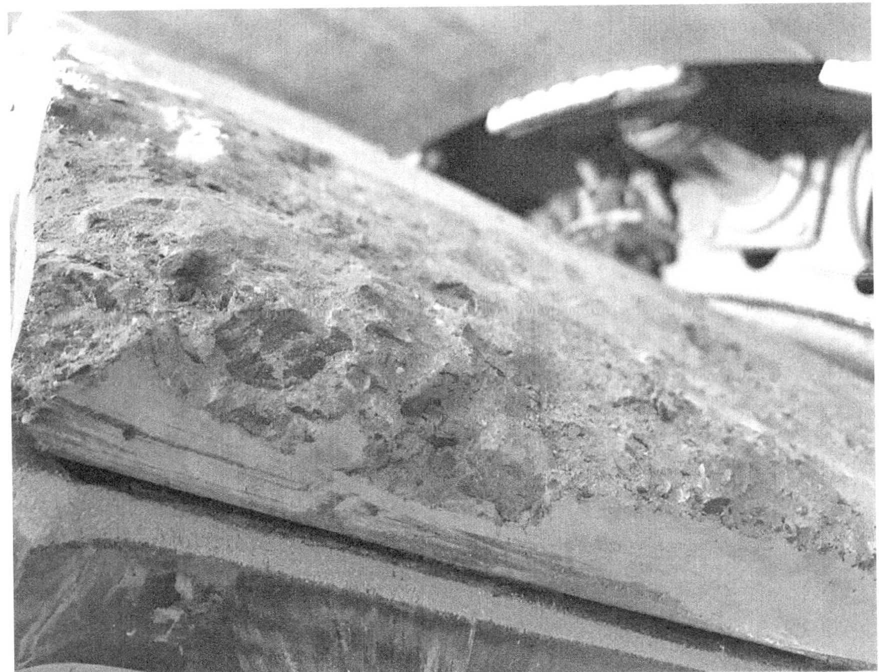

FIGURE 9.12 Uneven segment surfaces due to high slump value. (From the archive of the author, S. Acun.)

9.6 DESIGN ASPECTS OF SEGMENTAL TUNNEL LINING IN DIFFICULT GROUND CONDITIONS

Difficult ground conditions are an important issue in the design of precast segment design. However, it may be overlooked in some cases like high-speed railway tunnel T26, in Turkey leading to the collapse of the tunnel. A brief story about this tunnel will be given in this chapter to clarify the subject.

The excavation of T26 Tunnel started with New Austrian Tuneling Method (NATM) method in 2010. The tunnel is in the Northeast of Turkey. Due to geological problems, shear zones, fault zones, and low RQD values, the daily advance rates were very slow. The tunnel excavation method was changed to mechanical excavation using a 13.7 m diameter single shield TBM at the end of 2011. The tunnel was planned to pass through the predominately graphitic schist including frequent chlorite schist bands with sizes of some tens of meters and rarely through shear zones with a dimension of a few tens of meters. Graphitic schist is moderately weathered to fresh and very weak to weak and is extremely sheared along the foliation surfaces. Therefore, the schistose surface is polished and coated with clay. During the tunnel excavation phase collapses and jamming of the cutterhead started at chainage 216+ 300 m and continued thereafter. Segments started cracking in May 2012 and the tunnel started collapsing gradually damaging the TBM. For the sake of safety, all tunneling activities were stopped and the tunnel was abandoned thereafter. At the time of writing this chapter in January of 2022, the excavation of the tunnel was restarted by using conventional tunneling methods. Typical views of collapsed segments are given in Figures 9.13, 9.14, and 9.15 (Bilgin, 2016).

The case of T26 Tunnel was analyzed by Osgoui et al. (2021) in view of recommending some segment design criteria in such difficult geology. They commented on the significant increase of mechanized tunneling practically in all kinds of ground, including highly weathered rocks, severely jointed rocks, shear and fault zones, and in squeezing- and swelling-prone conditions. They also emphasized the need for further investigating the mechanical behavior and the damage mechanism of segmental tunnel lining in complex geologies. Different types of segment damage, due to over-stressing were reported in their research and as well as ring joint capacity performance. An effort was made to depict three scenarios in terms of in-situ stress compatibility and rock mass strength parameters for T26 Tunnel in such a way as to produce the realistic geotechnical model for each scenario. In the light of the analyses and real case representations given in this paper, the authors recommended well-calculated steel reinforcement to tackle the unfavorably asymmetrical loads on segmental lining resulting from adverse geological conditions and to increase the bending capacity of the segment.

9.7 CONCLUDING REMARKS

Segments are indispensable elements of mechanical excavation. However, in a tunneling project, the cost of segments may reach almost the majority of the cost of all equipment used including TBM. The high cost of the segments including safety issues necessitates a careful segment design, well-planned segment production, and

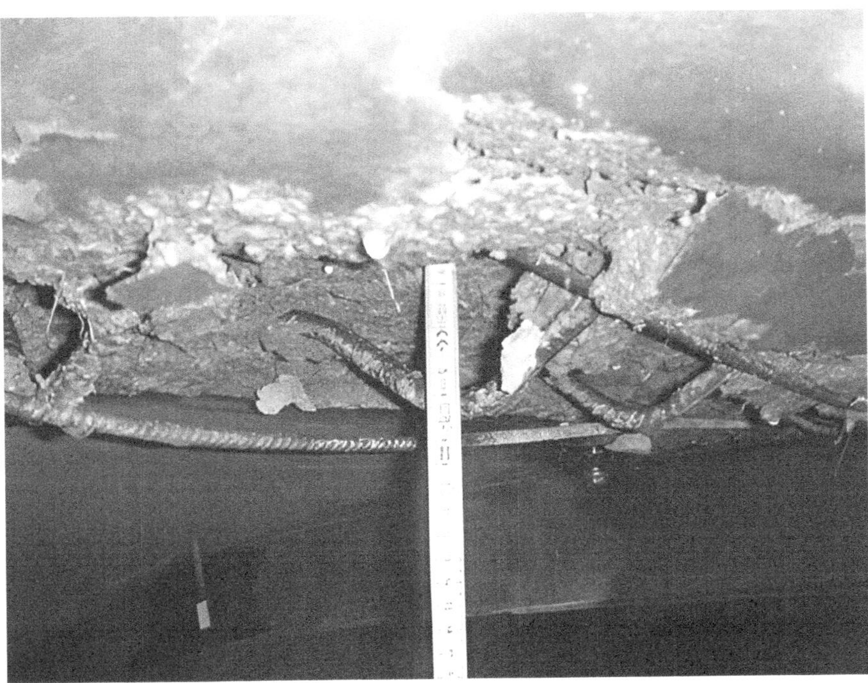

FIGURE 9.13 Segments failures in T26 Tunnel, Turkey. (From the archive of the author, N. Bilgin.)

FIGURE 9.14 Segments failures and steel arches used in T26 Tunnel, Turkey. (From the archive of the author, N. Bilgin.)

FIGURE 9.15   Segments failures in the floor of T26 Tunnel, Turkey. (From the archive of the author, N. Bilgin.)

a QM plan. The examples given in this chapter show that in segment production, the carousel plant needs less crew and less space compared to the stationary segment plant. Typical values for one of the most prestigious metro projects in Istanbul is 2 men/ring per shift in a carousel plant and 4.75 men/ring in a shift in a stationary plant. The ratio of indoor area to ring production per shift is 133.3 m^2/ring in a carousel segment plant and this ratio is 350 m^2/ring in a stationary plant. The other point emerging from this chapter is that a careful segment design is necessary for difficult ground conditions leading to segment failures and the closure of tunnels.

REFERENCES

Abbas, S., 2014. *Structural and Durability Performance of Precast Segmental Tunnel lining*, Ph.D. Thesis, The University of Western Ontario, p. 287.

Bilgin, N., 2016. An appraisal of TBM performances in Turkey in difficult ground conditions and some, *Tunnelling and Underground Space Technology* 57, pp. 265–276.

Cebeci, M., 2021. Istanbul-Airport metro project, presented by Murat Cebeci, the project director, *International Tunnelling Symposium Organized by Turkish Tunnelling Society in Istanbul*. Istanbul. https://tunnelder.org.tr

Daub, 2013. *German Tunnelling Committee (ITA-AITES) Recommendations for the Design, Production, and Installation of Segmental Rings*, p. 46.

Hariyanto, A.D., Kwan, H.P., Cheong, Y.W., 2005. Quality control in precast production, a case study on tunnel segment manufacture, *Dimensi, Teknik, Arsitrktur* 33(1), pp. 153–164.

Hurt, H., 2019. *Precast Concrete Segmental Liners, A Presentation of 89 Pages Prepared for ARUP*. https://oriolarnau.wordpress.com/home/segmental-tunnel-linings/.

ITA, 2019. *Report No 22, Guidelines for the Design of Segment Tunnel Linings*, April, p. 60.

Jiang, Y., 2019. Segment liner design and construction, *12th Annual Breakthroughs Short Course*, Atlanta, September 9–11.

Osgoui, R.R., Quaglio, G., Poli, A., Carrieri, G, 2021. Design aspects of segmental tunnel lining in difficult rock conditions, *Mechanics and Rock Engineering, from Theory to Practice*, in IOP Conference Series, *Earth and Environmental Science*, 833, p. 12, IOP Publishing. DOI:10.1088/1755-1315/833/1/012073

Praneshwari, R., 2019. Precast segmental lining for underground tunnels, *NBM&CW Infra Construction and Equipment Magazine*, October 2019.

Riechers, J., 2014. Customized state-of-the art,*TAI Journal India*, 3(1), pp. 15–22.

Shayanfar, M.A., Mahyar, P., Jafari, A., Mohtadinia, M., 2017. Classification of precast concrete segments damages during production and transportation in mechanized shield tunnels of Iran. *Civil Engineering Journal*, 3(6), June, pp. 412–426.

Simons, S., 2020. Integrated engineering for quality segmental linings, *TunnelTECH, Tunnel Talk*, 27 February 2020.

Singh, J., 2018. *Tunnel Lining Segment Production* (Lucknow Metro), CEPT University (Centre for Environmental Planning and Technology University), India, https://portfo lio.cept.ac.in/ft/project-5164-monsoon-2018/tunnel-lining-segment-production-luck now-metro-monsoon-2018-uc1914

10 TBM Spoil/Muck Management Plan

10.1 INTRODUCTION

Spoil management has been recognized as one of the key components of long tunnel construction and especially metro tunnels in urban areas. It should be planned before construction and organized well during the excavation process. Effective spoil management helps limit sound, dust, transport, and environmental emissions as well as is a cost-efficient way. Thalmann et al. (2013), summarized 20 years of spoil management experience in Switzerland with respect to the Lötschberg and Gotthard Base Tunnel projects in the mountains. A total of 16.5 million and 28.2 million tons were excavated from Lötschberg and Gotthard Base Tunnel projects, respectively. More than 30% of the excavated rocks were processed and used for concrete production. Pyeon (2016) in his report "Trend analysis of long tunnels worldwide", discussed problems emerging from long tunnels and the possibilities of using TBM spoil as aggregate. It should be borne in mind that environmental problems arising from TBM spoil are more complicated in highly populated urban areas, like Istanbul, where 20 million people live. Recently Cebeci (2021) and Öztürk (2021) reported that the spoil coming from only two metro projects for Istanbul's new airport was around 11.106 m^3 per three years of construction time. However, it is reported that in 2021 the total solid waste generated from excavation and demolishing activities in Istanbul was around 40.106 m^3. Figure 10.1 shows the diversity of solid waste generation in big cities like Istanbul.

Construction and demolition waste (CDW) is the largest waste stream in the European Union (EU) by volume, with about 850 million tons produced annually. Belgium is among the leading European countries for the recycling of CDW. Germany is the first EU country by volume, with about 68 million tons recycled every year, but proportionally the Netherlands is the best, with about 90% of material recycled (Sforza, 2018). The environmental management plan for CDW differs from that of the environmental management plant for TBM spoil. For this reason, this chapter is written only with the intention of describing the characterization, the potential use of the TBM spoils in order to create an environmental management plant and also for trying to find an answer to the question "Is TBM spoil a problem or a valuable asset?"

The circular economy is a trend in response to the inefficient management of resources in the traditional linear model and promotes greater resource productivity,

DOI: 10.1201/9781003358978-10

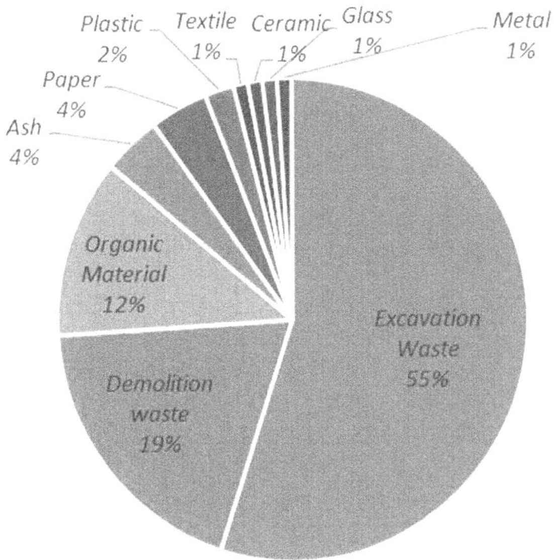

FIGURE 10.1 Solid waste generation of Istanbul. (Ocak, 2009.)

aiming to reduce waste and avoid pollution by design or intention (Michelinia et al., 2017: Romero and Rossi, 2017; Korhonen, 2018). In other words, the utilization of excavated rock and soil has gained worldwide significance in a circular economy based on the principle "Reduce, Reuse and Recycle" (Haas et al., 2021). Passing from linear to circular economy is well illustrated and explained by Kivivirta (2021).

In several countries, government authorities started pushing hard for the contractors to follow the basic rules of the circular economy. As reported by Hale et al. (2021), in 2020, the EC (European Commission) published a report on "Circular Economy Action plan for a cleaner and more competitive Europe". Within this report, a new strategy for a sustainable built environment is outlined and one goal is "promoting initiatives to reduce soil sealing, rehabilitate abandoned or contaminated fields and increase the safe, sustainable and circular use of excavated materials". However, according to these authors, there are barriers to the reuse of excavated soil, which are mainly as follows.

a. Regulatory barriers. They are defined as those that arise as a result of existing (or lack of) regulations from environmental authorities and other regulative bodies.

b. Organizational barriers. The main organizational barrier to the reuse of excavated soil is that the project planning process, in its current form, simply does not consider reuse of soil in most countries.

c. Logistical and economic barriers. A problem arises related to the storage space needed for excavated soils in the interim before subsequent use. In many construction projects, especially in densely populated cities, on-site space is extremely limited, and excavated soils must be removed to temporary storage sites.

d. Material quality barriers. Excavated soils can be contaminated, their degree of contamination must be well characterized and assessed as the first step for reuse.

10.2 CHARACTERIZATION OF THE TBM SPOIL, PROBLEMS, AND THE NEED FOR NEW TECHNOLOGIES

For efficient planning of a TBM spoil management, it is strictly necessary to know the characteristics of the muck such as: petrography, physical and mechanical properties, shape, granulometry, and contamination from chemical agents used. Each of these properties will affect the reuse of the TBM spoil. However, in complex geologies, such as encountered during a hard rock TBM drive in Beykoz/İstanbul (Figure 10.2) the spoil may have completely different characteristics within different parts of the tunnel route, as seen in Figures 10.3, 10.4, and 10.5 (Güclücan et al. 2008; Bilgin et al., 2016; Bilgin, 2016). The diversity of the TBM spoil such as in the Beykoz tunnel creates problems in making an efficient spoil management plan. Limestone chips were screened and sorted by size and used as concrete material during the construction of the tunnel. However, the TBM spoils seen in Figures 10.3, 10.4, and 10.5 are accumulated in the tunneling site area and later transported to private lands hired by the contractor.

Diversities of TBM spoils as shown above, including environmental and legal restrictions within the concept of circular economy, led the academicians and industry practitioners to carry out advanced technological research studies on the innovation in spoil management. A good example of these research studies is the DRAGON project sponsored within the EU's Seventh Programme for research, technological development and demonstration under the grant agreement No. 308389 with a budget of more than 4.10 million euros. The main technological developments within the DRAGON project include fully automatic processes for rapid detection of usable materials, recycling the materials immediately on the TBM as outlined in Figure 10.6. Although the main future of this promising technology for efficient TBM spoil management was published by Galler (2015, 2019), no further information is available on its practical use in current projects and we believe that further efforts are needed to put this project alive.

10.3 POTENTIAL USE OF TBM MUCK FOR CONSTRUCTION APPLICATIONS

10.3.1 Spoil/Muck from Hard Rock TBMs

TBM muck must meet certain physical and chemical properties and have economic benefits to be utilized as a construction material. The properties of TBM muck are highly dependent on the geology of the area from which it has been extracted as well as the TBM technology used for excavation, such as hard rock TBM, EPB or slurry TBM. Hence, the excavated material from each TBM project should be studied

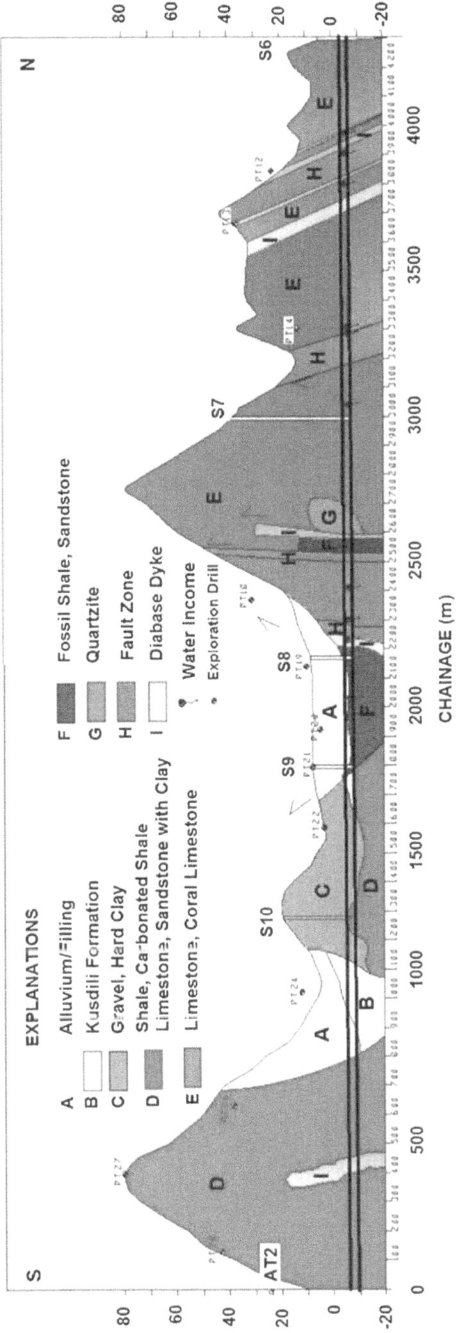

FIGURE 10.2 Geological profile of the Beykoz/Istanbul sewerage tunnel. (Bilgin, 2016.)

FIGURE 10.3   Diabase dyke in Beykoz/İstanbul Tunnel. (From the archive of the author, N. Bilgin.)

FIGURE 10.4   Marn particles in Beykoz/İstanbul Tunnel. (From the archive of the author, N. Bilgin.)

FIGURE 10.5 Muddy TBM spoil from Beykoz Tunnel/İstanbul. (From the archive of the author, N. Bilgin.)

separately to determine its suitability as a construction material and to decide on the processing requirements (Taqa et al., 2021).

In hard rock TBM drives, in most cases cuttings appear to have approximately the correct average size for some applications where aggregate is a common requirement. If they are suitable in other respects, cost savings can be realized in tunnel construction. A review of standard construction aggregate specifications indicates that hard rock TBM muck would be suitable for several construction applications with minimum processing, such as for structural concrete, and road pavement (Gertsch, 2000; Riviera, 2013). A useful guide for characterization of TBM muck for aggregates, for road works/embankments, as raw materials for industry, environmental or land reclamation and backfilling is discussed by Oggeri and Ronco (2010) and summarized by them (2017) as given in Table 10.1. Utilizing excavated rock material from tunnel boring machines is summarized also by Berdal et al. (2018).

TBM spoil may be used mostly as aggregate for concrete and asphalt pavement and also as construction aggregates and ballast. Excavated tunnel materials from tunnels in high strength rocks, represent excellent and valuable aggregate source for different purposes. Some basic requirements are well described in ITA report No. 2, 2019.

Concrete aggregates: Basic requirements for concrete aggregates are given in European Standard EN12620:2002+A1. The requirements are related to the geometry (size, grading, shape), content and quality of fines, physical property and strength, (resistance against crushing, wearing, flakiness, density, water absorption, durability) and chemical requirements (chlorides, sulphate content, carbonate, mica content, etc.)

Aggregates for bituminous materials and road pavements, airport, and other traffic areas: Basic requirements are given in European Standard EN13043. The requirements are related to grain size, shape and grading, content of fines, rock strength properties, flakiness, resistance against crushing (Los Angeles value, resistance against wearing, etc.).

Aggregate for railway ballast: The requirements are given in European Standard EN12620:2002+A1. As for concrete aggregates, the requirements are related to grain

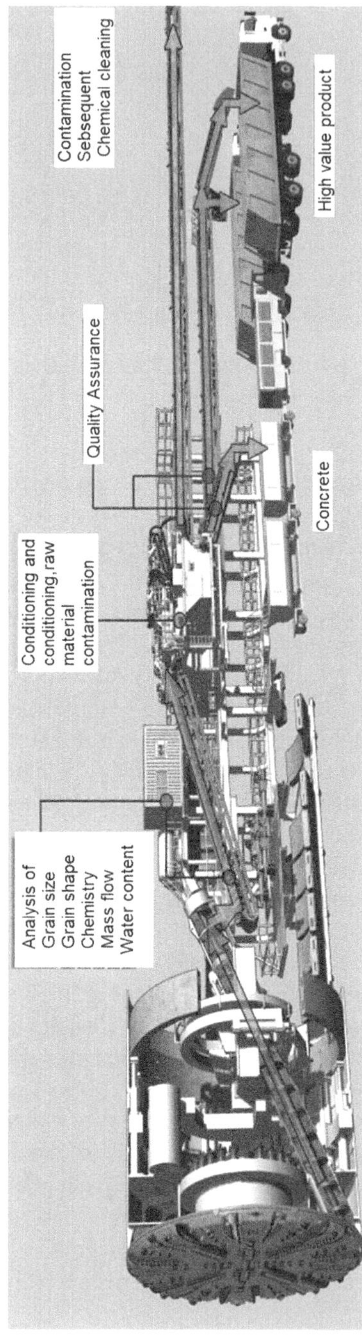

FIGURE 10.6   The concept of the DRAGON European Project. (Galler, 2015.)

TABLE 10.1
The Classification of TBM Muck for the Use in Construction Industry and Relevant Test Parameters

Parameter	Aggregates for Constructions	Road Works/ Embankments	Raw Material for Industry	Environmental or Land Reclamation	Back Filling
Deformability modulus; coefficient of consolidation; compressibility indices; bearing capacity	X	XXX	O	XX	XX
Uniaxial compressive strength; shear strength; angle of friction; cohesion	XX	XXX	XX	XX	X
Spoil particles structure, voids ratio; compaction; unit weight	X	XXX	X	XX	XX
Durability, resistance to wear, to fragmentation and frost; alkali aggregate reaction	XXX	XX	XXX	X	X
Leaching properties	XX	XX	XX	XXX	XX
Ore grade and stone value	XX	X	XXX	O	O
Particle size distribution and morphology, presence of fines; petrography; shape of elements	XXX	XXX	X	X	X
Water interactions water content; consistency indices; hydraulic conductivity	XX	XXX	X	XX	XX

Source: Oggeri and Ronco (2010).

Note: XXX: relevant; XX: important; X: minor concern; O: not applicable.

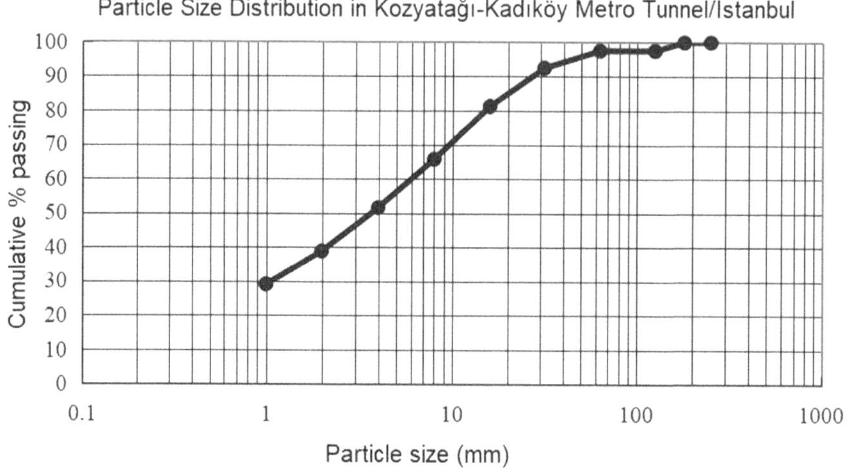

FIGURE 10.7   Particle size distribution of the TBM muck in Kozyatağı-Kadıköy Metro tunnel. (From the archive of the author, N. Bilgin.)

size, shape and grading, content of fine, rock strength properties, flakiness, resistance against crushing (Los Angeles value and SZRB value), resistance against wearing (Micro Deval coefficient), resistance against freezing and thawing, water absorption, etc.

TBM spoil may be used also for land reclamation, embankments, foundations and landfill, for earth and rock-fill dams, for erosion, shore and slope protection, for rockfall defense, and for backfill to borrow areas and landscape rehabilitation (Rahimzadeh, 2018).

Particle size distribution of the muck is given in Figure 10.7 representing Karatal limestone having a mean compressive strength of 65 MPa and is obtained from Kozyatağı-Kadıköy Metro tunnel. This muck is used partly as aggregate in construction industry and partly in embankment and in landfill after screening and processing.

A typical use case for hard rock TBM muck is illustrated in the Gotthard base tunnel where approximately the following distribution was achieved, 20% of the tunnel muck was used for embankments related to the project, 30% as aggregate for concrete production including lining segments, 44% sold to third parties for environmental restoration (Thalmann et al. 2013).

Earth and rock fill dams are constructed to create reservoirs or to increase the volume in natural lakes, and thus to increase the volume or water supply. The muck is also used as filling material in an abandoned quarry pit designed for water storage (Tokgöz, 2013).

10.3.2   Muck/Spoil from Soft Ground TBM Tunnels

Considerable attention needs to be paid to the handling, treatment, and disposal of the tunnel spoil generated by soft ground tunnels. Unlike rock spoil, soft ground tunnel

FIGURE 10.8 Spoil handling in Gayrettepe-Istanbul Metro Project, Hastal station, photograph by the author, S. Acun.

spoil more frequently has few uses and is typically more difficult to dispose of. Some spoil requires secondary treatment such as drying and chemical additives to make it suitable for using such as in land fill. Only small quantities of this spoil may be used in engineering activities. Spoil from soft ground tunneling may be considered contaminated and hazardous due to the bentonite and chemical mixing coming from the foam, bentonite, etc. Figure 10.8 gives a general view of spoil handling in Gayrettepe-Istanbul Metro Project, Hastal station, where EPB-TBMs were used. The reader should notice the drying and leachate pond, with lorries around for transporting spoil material. Transporting of such TBM spoil needs careful planning of spoil management in urban areas due to traffic and environmental restrictions. The transport and the storage of the spoil may affect the economy of mechanized tunneling operation tremendously. A typical example of the rational planning of the spoils in Halkalı-New Istanbul airport is given in Figure 10.9. A special place is organized next to the center where the tunneling activities are running. Thereafter, recreation and cultivation operations are carried out as seen in Figure 10.10. For this project, 50,000 trees were planted (Öztürk, 2022).

10.4 PLANNING OF THE MANAGEMENT ON TBM SOIL WITHIN THE FRAME OF CIRCULAR ECONOMY

A literature review on sustainable management of excavated soil and rock in urban areas is given by Magnusson et al. (2015). According to Schwartzentruber and

FIGURE 10.9   The storage of the TBM spoil in Halkalı- Istanbul Airport. (Öztürk, 2022.)

FIGURE 10.10   Rehabilitation of the spoil storage area in Halkalı-İstanbul Metro Project area. (Öztürk, 2022.)

TABLE 10.2
Basic Concepts and Description of the Management Plan of Excavated Materials

Item	Action
Baseline and environmental setting	Potential sensitive receptors, meteorology, hydrogeology, drainage, topography, geology, petrography, minerology, physical and mechanical properties, abrasivity
Tunnel spoil management and disposal	TBM spoil characteristics, standard tests for reuse of TBM spoil as aggregates, landfill, etc. TBM spoil water sampling, leachate reuse criteria
Tunnel site facility	Design holding and drying ponds, leachate sedimentation pond, slope stability, internal traffic routes
Planning, logistics, and facilities	Storage capacities and location, reprocessing and transportation, design of reprocessing units, time scheduling
Environmental management program	Environmental training, dust suppression, pollution incident plan, complaints, periodic performance audit
Environmental risk assessment	Source, pathway and receptors, risk assessment methodology, ground and surface water risk assessment, land soil risk assessment, noise control and management strategies
Territorial needs and economy	Needs and usages (flow-analysis), identification of potential valorization sectors and sites, securing supplies (quantities and alternative option)
Closure measurements and monitoring	Groundwater, surface water and soil monitoring, air quality and dust suppression monitoring, surface reinstatement and revegetation

Source: Compiled from CETU (2016), Schwartzentruber and Robert (2018), and Callege (2021).

Robert (2019), CETU (Tunnel Study Centre in France) initiated a research program on the management of excavated materials during underground works. The first step consisted in publishing an information document to increase awareness about the management of excavated materials and to make sure that they would be considered as a natural resource (CETU, 2016). Table 10.2 gives the basic concepts and description of the management plan of the excavated material.

In light of the published reports and papers, the following items are considered in the material used plan: maximum distance from the construction feature to the material storage area, minimum distance to existing residential area or commercial buildings, landowners' concerns and preferences, locations of the storage areas for reusable tunnel material, chemical characterization of dumped material, etc.

For initiation, documentation of on-site settings with the legislative framework and the approval of the requirements are necessary. The second step is to study environmental settings like; site ownership, potential sensitive receptors, meteorology, geology, geotechnical properties of the rock formation, hydrogeology, topography, drainage, background air quality, and noise. The further step in the planning of the management of TBM spoil is tunnel spoil management and disposal, which will include, TBM spoil characteristics, standard tests for reuse of TBM spoil in different industrial applications, TBM spoils water sampling, leachate reuse possibilities, and methodology for closing the loop of the circular economy. Environmental risk assessment, environmental management program, and closure measurement and monitoring are the final stages of the planning of the management of TBM spoil. The readers are advised to consult references Schwartzentruber et al, (2019) and Callege (2021), for more detailed information on the subject.

10.5 CONCLUDING REMARKS

A spoil management plan for TBM spoil is a necessity in any mechanized tunnel excavation due to environmental legislative restrictions and it must close the loop of the circular economy. If the TMB spoil meets some standard requirements, it may be used in different construction applications with minimum processing, such as for structural concrete, road pavement, aggregates, road works/embankments, raw materials for industry, backfilling, and as rock filling material in abandoned quarries for water storage. At the end of this chapter, all the requirements for efficient planning of the management of TBM spoil are given with the aid of Table 10.2

REFERENCES

Berdal, T.C., Jakobsen, P.D., Jacobsen, S., 2018. Utilizing excavated rock material from tunnel boring machines, in *SynerCrete'18 International Conference on Interdisciplinary Approaches for Cement-based Materials and Structural Concrete*, 24–26 October 2018, Funchal, Madeira Island, Portugal.

Bilgin, N., 2016. An appraisal of TBM performances in Turkey in difficult ground conditions and some recommendations. *Tunnelling and Underground Space Technology*, 57 (2016), pp. 265–276.

Bilgin, N., Çopur, H., Balcı, C., 2016. *TBM Excavation in Difficult Ground Conditions, Case Studies from Turkey*, Wiley, Ernst and Sohn, p. 336.

Callege, D., 2021. *Tunnel Boring Machine Spoil Environmental Management Plan*, April, Western Soil Treatment Pty Ltd, p. 168.

Cebeci, M., 2021. Istanbul-Airport Metro project, presentation made by Murat Cebeci, the project director, in *International Tunnelling Symposium Organized by Turkish Tunnelling Society*, Istanbul.

CETU, 2016. Matériaux géologiques naturels excavés en travaux souterrains – Spécificités, scénarios de gestion et rôle des acteurs. *Document d'information CETU*, March 2016, p. 31.

Galler, R., 2015. Development of resource-efficient tunnelling technologies – results of the European research project DRAGON. *Geomechanics and Tunnelling*, 8(4). doi:10.1002/geot.201500017

Galler, R., 2019. European research results on resource efficient tunnelling-Tunnel excavation material – waste or valuable mineral resource?, in *Tunnels and Underground Cities: Engineering and Innovation Meet Archaeology, Architecture and Art* – D. Peila, G. Viggiani, & T. Celestino (eds), Taylor & Francis Group, London.

Gertsch, L., Fjeld, A., Nilsen, B., Gertsch, R., 2000, Use of TBM muck as construction material. *Tunnelling and Underground Space Technology*, 16(4), pp. 372–402.

Güclücan, Z., Meric, S., Algan, M., Palakci, Y., Bilgin, N., Bilgin, A.R., Balci, C., Tumac, D., 2008. The use of a TBM in difficult ground conditions in Beykoz-KavacikSewerage tunnel, in *Proceedings of the World Tunnelling Congress, Underground Facilities for Better Environment and Safety*, Agra, India, pp. 1630–1638.

Hale, S.E., Roque, A.J., Okkenhaug, G., Sørmo, E., Lenoir, T., Carlsson, C., Kupryianchyk, D., Peter Flyhammar, P., Bojan Žlender, B., 2021. The Reuse of excavated soils from construction and demolition projects: Limitations and possibilities. *Sustainability*, 13, p. 6083. doi.10.3390/su13116083

Haas, M., Mongeard, L., Ulrici, L., Laetitia D'Aloïa, L., Cherrey, A., Robert Galler, R., Benedikt, M., 2021. Applicability of excavated rock material: A European technical review implying opportunities for future tunnelling projects, *Journal of Cleaner Production*, 315(2021), p. 128049. doi.10.1016/j.jclepro.2021.128049

ITA report No. 2, 2019. *Handling, Disposal of Tunnel Spoil Materials*, April, p. 60.

Kivivirta, M.M., 2021. *Management of Excavated Soil in Infrastructure Construction Using Multi-Criteria Decision Analysis*, Master's Thesis, Aalto University, p. 46.

Korhonen, J., Honkasalo, A., Seppälä, J., 2018. Circular economy: The concept and its limitations. *Ecology economy and society*, 143, pp. 37–46. doi:10.1016/j.ecolecon.2017.06.041

Magnusson, S., Lundberg, K., Svedberg, B., Knutsson, S., 2015. Sustainable management of excavated soil and rock in urban areas, a literature review. *Journal of Cleaner Production*, pp. 1–8. doi:10.1016/j.jclepro.2015.01.010

Michelinia, G., Moraesa, R.N., Cunhab, R.N., Costaa, J.M.H., Omettoa, A.R, 2017, From linear to circular economy: PSS conducting the transition, in *The 9th CIRP IPSS Conference: Circular Perspectives on Product/Service-Systems*, *Procedia CIRP* 64, pp. 2–6. doi:10.1016/j.procir.2017.03.012

Ocak, I., 2009. Environmental problems caused by Istanbul subway excavation and suggestions for remediation. *Environmental Geology*, 58, pp. 1557–1566. doi:10.1007/s00254-008-1662-9

Oggeri, C., Fenoglio, T.M., 2017.Tunnelling muck classification: definition and application, in *Proceedings of the World Tunnel Congress 2017 – Surface Challenges – Underground Solutions*, Bergen, Norway.

Oggeri, C., Ronco, C., 2010. Investigation and test for reuse of muck in tunnelling, in *Proc. WTC 2010, Tunnel Vision Towards 2000*, TAC, Vancouver.

Öztürk, M., 2021. Metro project for Halkalı-Istanbul-Airport metro project, presentation made by Murat Cebeci, the project director, in *International Tunnelling Symposium Organized by Turkish Tunnelling Society*, Istanbul, 24–25 December.

Öztürk, M., 2022. Istanbul-Airport Metro project, presented by Mustafa Öztürk, the project director, in International Tunnelling Symposium organized by Turkish Tunnelling Society, Istanbul, 28–29 November.

Pyeon, J.H., 2016. *Trend Analysis of Long Tunnels Worldwide*, March 2016, Mineta Transportation Institute, Report WP 12–09, p. 62

Rahimzadeh, A., Tang, W.C., Sher, W., Davis, P., 2018. Management of excavated material in infrastructure construction- A critical review of literature, January 2018, in *Conference: International Conference on Architecture and Civil Engineering*, Sydney, Australia.

Riviera, P.P., Bellopede, R., Marini, P., Bassani, M., 2013. Performance-based re-use of tunnel muck as granular material for subgrade. *Tunnelling and Underground Space Technology*, 40(2014), pp. 160–173. doi:10.1016/j.tust.2013.10.002

Romero, D., Rossi, M., 2017. Towards circular lean product-service systems, in *The 9th CIRP IPSS Conference: Circular Perspectives on Product/Service-Systems*, *Procedia CIRP*, 64, pp. 13–18. doi:10.1016/j.procir.2017.03.133

Schwartzentruber, L.D., Robert, F., 2019. Management and use of materials excavated during underground works, in *Tunnels and Underground Cities: Engineering and Innovation Meet Archaeology, Architecture and Art* – D. Peila, G. Viggiani, & T. Celestino (eds), Taylor & Francis Group, London.

Sforza, M., 2018. Recycled construction waste: building a more sustainable future, *European Research Media Center, youris.com*, 9 October.

Thalmann, C., Petitat, M., Kruse, M., Pagani, L., Weber, B., 2013. Spoil Management: Curse or Blessing? Looking back on 20 years of experience, in *Underground. The Way to the Future*, G. Anagnostou & H., Ehrbar (eds), Routlege, Taylor and Francis Group, Oxfordshire, pp. 1659–666.

Taqa, A.A., Al-Ansari, M., Taha, R., Senouci, A., Al-Marwani, H.A., Al-Zubi, G.M., Mohsen, M.O., 2021. Characterization of TBM muck for construction applications. *Applied Science MDPI*, 11(8623), p. 1–17. doi:10.3390/app11188623

Tokgöz, N., 2013. Use of TBM excavated materials as rock filling material in an abandoned quarry pit designed for water storage. *Engineering Geology*, 153(2013), pp. 152–162. doi:10.1016/j.enggeo.2012.11.007

11 Management of Contaminated Ground During and After TBM Drives

11.1 INTRODUCTION

Contaminated land refers to the ground and, in many instances, groundwater where concentrations of hazardous chemicals exceed those specified in regulations determined by health and public authorities. Contamination occurs in a variety of forms, but mainly from soil conditioning agents in TBM excavation, excavating geologic formations containing asbestos, and organic compounds such as petroleum hydrocarbons (Langmaack 2017). The worst may be contamination by carcinogenic chemicals known as per- and polyfluorinated alkyl substances (PFAS). A typical example of this is the West Gate project in Melbourne, causing an extra overburden of $3.3 billion to the project to remove the contaminated ground, to store it somewhere isolated from human beings, fauna, and flora. All these negative aspects will be treated in this chapter by giving examples from different TBM projects. There are also naturally occurring acid sulfate soils and rock. If disturbed during excavation, these soils and rock can oxidize and produce sulfuric acid.

11.2 POSSIBLE CONTAMINANTS

TBM additives, in soil conditioning, may be one of the reasons for the ground contamination resulting in dramatic environmental problems. A typical example is The Hallandsås Tunnel in Sweden (1997). Using over dosage of acrylamide in the grout injection polluted a total of 29 wells in the area, and the cows that had drunk seepage water became paralyzed and eventually died. Another example is Kishanganga Hydro Electric Project Head Race Tunnel, India, ending with the death of the cows with the revolt of the farmers and the intervention of the army in 2014. Due to this reason, it is stickily advised to use environmentally friendly solutions in soil conditioning and follow the latest standards restricting unfriendly solutions (Sposetti 2021).

Another problem may arise from the excavation, treatment, and storage of the ground already contaminated with carcinogenic chemicals known as PFAS. There may be a shortage of sites for the storage of such spoil, as occurred in the West Gate project in Melbourne. Two TBMs of 15.6 m diameter came to the side. As of the date of writing this chapter, the project could not start yet due to the dispute between the job owner and the contractor. However, at the revision of this chapter in an article published in tunnels and tunneling on the 3 March 2022 it is reported that "The largest TBM in the Southern Hemisphere has started excavating the West Gate Tunnel in Melbourne, Australia". The contractors claim that there is a "Force Majeure Termination Event" since the discovery of the contaminated soil was discovered after the contract was signed. It is expected that the $6.7 billion project will need $3.3 billion extra to finish the project. It is indicated that the blowout could be even bigger because the builder's claims are higher (Rowland 2020; Willingham 2020; Edwards 2021) www.tunnelsonline.info/news/australias-largest-tbm-starts-on-west-gate-tunnel-9524299.

Istanbul faces frequent earthquakes, which loosen the pipe connections of underground petroleum tanks. During underground works, it is sometimes faced with petroleum-contaminated ground, which causes a tremendous fire risk during tunnel excavation. Fuel-oil stations are usually encountered when constructing metro lines in densely populated urbanized areas, which require measures/precautions at a very high level. The hydrocarbon pollution mechanism is shown in Figure 11.1. Typically, hydrocarbons form a layer that floats on the water table in a light non-aqueous phase (LNAPL). Due to its higher concentration of contaminants and lack of dissolution, the LNAPL presents a larger threat to construction workers. However, where the tunnel crown passes at a distance of more than 5 m from the phreatic surface level, the groundwater acts as a buffer zone, eventually protecting the tunnel from contaminants (Stewart 2019). Notwithstanding, dissolved contaminants and hydrocarbon vapors may also pose a probable threat to the worker's health and a possible explosion risk depending on the environmental conditions. In particular, petrol components such as benzene, which is soluble in water, may change back again into vapor when extracted

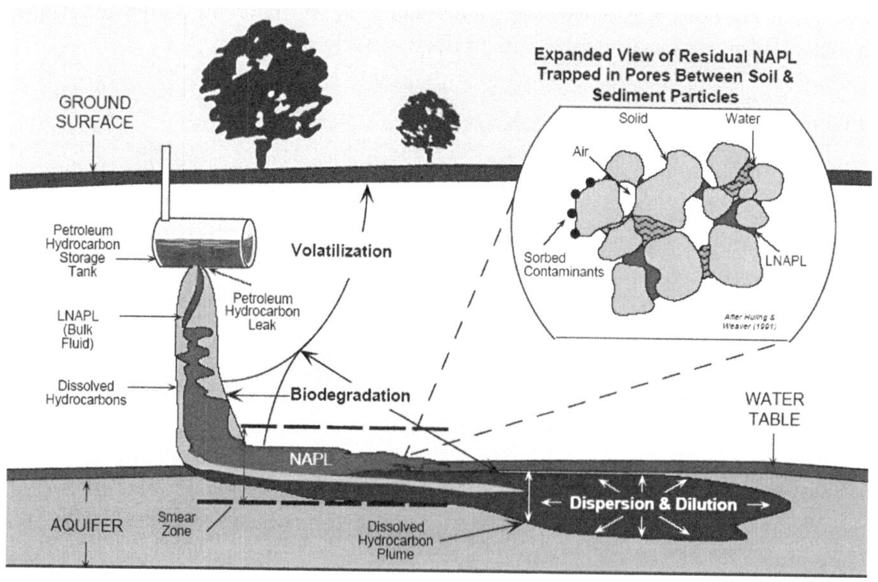

FIGURE 11.1   Typical hydrocarbon contamination mechanism. (Steward 2019.)

with the screw conveyor of the TBM and reach atmospheric pressure, becoming potentially dangerous to human health (Yazici et al. 2019; Bilgin et al. 2021).

Typical examples of hydrocarbon-contaminated ground in Istanbul are the Uskudar–Umraniye–Cekmekoy metro and Bakirkoy–Kirazli metro projects. These two examples will be treated in detail in Sections 11.5.1 and 11.5.1 when explaining the management of tunnel fire risks during TBM driving in the hydrocarbon-contaminated ground.

11.3 WHAT IS SOIL CONDITIONING AND WHY?

EPB-TBMs have successfully been adopted for urban tunneling in recent years in very different ground conditions and, at present, they can be considered the most commonly used mechanized tunneling technology in soft ground in urban areas. This increase in the number of tunnels excavated by EPB machines has been possible due to the more effective use of additives to condition the soft ground. Soil conditioning is obtained through the injection of foam, polymers, water ahead of the tunnel face and into the bulk chamber, and along the screw conveyor. The aim of this procedure is to transform the soil into a plastic medium that will be able to transmit the pressure properly in the excavation chamber and along the screw conveyor to reduce the permeability of the ground. This consequently prevents water inflows, seepage destabilizing forces, and consolidation-related surface settlement; reduces frictional forces, tool forces, and mechanical wear; and decreases the required torque of the cutting head and of the screw conveyor. Using anti-clogging agents also avoids the adhesion effects of sticky clayey soils (Vinai et al. 2008). After Peila and Picchio (2011), the most used additives for soil conditioning are foam, polymers, anti-amalgams, abrasion preventers, bentonite, and filler. Surfactants are delivered to the site as concentrates. These are mixed with water and compressed air in a specific ratio to create a foam. Polymers improve the foam stability and the consistency of the spoil. To be more acquainted with the subject, the readers are advised to follow the concepts defined in EFNARC (2005) which are summarized in the following.

Some parameters, such as foam expansion ratio (FER), foam injection ratio (FIR), and surfactant (foam) concentration (CF), determine the success of a conditioning process. FER in % is defined as the ratio of the foam volume used to the unit volume of the excavated ground and estimated by FER = V foam/V foam solution. V foam is the volume of foam at working pressure and V is the foam solution volume of foaming solution. The FER should be at 5–30. The higher the FER, the drier the generated tunnel foam will be. The wetter the soil, the drier the tunnel foam should be, and vice versa.

FIR is defined as FIR = 100 × V foam/V soil. V foam is the volume of foam at working pressure and V soil is the volume of in situ soil to be excavated. The FIR can be at 10%–80%, in most cases around 30%–60%. To determine the best FIR value, laboratory tests have to be carried out. The water content of the soil or the amount of injected water plays an important role.

CF is the concentration of surfactant agent in water and is calculated as CF = 100 × m surfactant/m foam solution, m being surfactant mass of surfactant in foaming solution

TABLE 11.1
Foam Types for EPBM Relative to Different Soils

	Foam Types				Polymer Additives
Soil	A	B	C	FIR	
Clay	↕			30–80	Anti-clogging polymer
Sandy clay-silt		↕		40–60	Anti-clogging polymer
Sand clayey-silt				20–40	Polymer for consistency control
Sand			↕	30–40	Polymer for cohesiveness and consistency control
Clayey gravels				25–50	Polymer for cohesiveness and consistency control
Sandy gravels				30–60	Polymer for cohesiveness and consistency control

Source: EFNARC (2005).

Note: FIR values are indicative only.

and m foam being mass of foaming solution. The concentration of foaming solution CF is typically in the range of 0.5%–5.0% but should follow the manufacturer's recommendations. This concentration strongly depends on the amount of water that is injected or which is already present in the soil and also affects the activity and stability of the used tunnel.

The foam type chosen should match the properties of the soil to be excavated, as seen in Table 11.1. Foam type A has a high dispersing capacity (breaking clay bonds) and good coating capacity (reducing swelling effects). Foam type B is a general-purpose foam having medium stability. Foam type C has high stability and anti-segregation properties to develop and maintain cohesive soil as impermeable as possible.

11.4   GENERAL RISK CHARACTERISTICS OF SOIL CONDITIONING/TBM ADDITIVES

Chemicals used during TBM drives encounter underground water and remain in the excavated spoil so they must be environmentally friendly, not hazardous. Therefore, the selection of soil conditioning agents must be based on the combined influences of technical efficiency, eco-toxicity, and biodegradability.

The eco-toxicity of chemical products, and foaming agents, is determined in accordance to meet the environmental needs of the international community (OECD 203, 2000). The OECD is an international organization for economic cooperation and development that works on establishing international standards for a range of social, economic, and environmental challenges (OCED 301). The guidelines rank the level of risk that a chemical product poses, to water and land organisms, into three categories: WGK 1 as low risk, WGK 2 as medium risk, and WGK 3 as high risk. The level of risk for eco-toxicity is determined from the results of different tests outlined in the following OECD guidelines which measure the effect of the chemical on various species: for toxicity to fish (LC50)–OECD203, for toxicity to Plankton

(EC50)–OECD202, for toxicity to algae (IC50)–OECD201, for toxicity to mammals (LD50)–OECD420.

Biodegradability is determined by measuring the addition of organic material to the soil as a result of conditioning with foaming agents and the length of time that the added organic material remains in the soil (OECD301). The two measures for making this evaluation are: the total amount of oxygen required for complete degradation of the organic material added to a given system and the amount of dissolved oxygen demanded by aerobic organisms to break down organic compounds.

Environmental management plan before TBM mining
The highest acceptable range of concentration limits of chemicals in the ground should be determined by soil conditioning design and specs. Chemical suppliers need to be involved in this stage. In the early stage of the muck management plan, the time muck can leave the job site, identification of approved dump areas for muck disposal and potential plans for muck reuse (i.e., backfill, construction) should be included in this plan. This plan will be developed later and implemented by future contracting teams.

Environmental muck management and risk assessment
Following the completion of the project, after a certain time depending on the estimated velocity of migration of polymers in the ground, a review campaign of the groundwater should be carried out, to measure the following: pH, total dissolved solids in mg/L, total suspended solids in mg/L, the total amount of oxygen required for complete degradation of the organic material added to a given in mg/L, the amount of dissolved oxygen demanded by aerobic organisms to breakdown organic in mg/L, oil and greases in mg/L. Job owners, contractors, and regional authorities should take part in long-term muck management.

11.4.1 ECOLOGICAL AND TOXICOLOGICAL PROPERTIES OF SOIL CONDITIONING AGENTS

Soil conditioning agents not only have to fulfill performance criteria but also toxicological and eco-toxicological criteria. As a consequence, the possible impact on the surrounding environment plays an important role in the choice of soil conditioning additives and is one of the exclusion criteria. To determine the possible risk of a product, a risk assessment study should be carried out. The most important factor in this study is the risk evaluation for human beings and the environment. It is mainly determined by the following four points:

1 The amount of substance entering the environment
2 The chemical and physical properties of a substance that determine the distribution in the environment. In most cases, this is the leachability of groundwater. In addition, bioaccumulation has also to be taken into consideration
3 The toxicity of a substance to the environment, respectively the toxicity towards aquatic organisms and humans
4 The elimination process (degradation). Organic substances can be degraded as follows:

Toxic effects depend on the amount of a substance that is available to the organism. Toxicity tests carried out in the laboratory are used to predict the so-called safe concentrations at which no negative impact on the organisms is expected. Bioaccumulation is a process by which organisms concentrate chemicals within themselves. This can result either from their food or directly from the surrounding environment. Biodegradation is the breakdown of an organic substance by the action of microorganisms. Before degrading completely to water and CO2, substances may degrade into smaller intermediates. Persistence is the ability of substances to resist degradation.

11.4.2 Suitable Soil Conditioning Products

Suitable soil conditioning products should only be those which show the desired functional properties and at the same time are as safe as possible for the workers and the environment. This implicates a judgment of the acute aquatic toxicity, potential for bioaccumulation, biodegradation, and chronic aquatic toxicity by risk assessments.

The most sensitive area is acute aquatic toxicity. Tests should be done according to OECD Guidelines 201 to 203. The ecological properties of a product are judged by biodegradation data, using OECD guidelines with a defined number of starting bacteria. Generally, soil conditioning products shall be either readily biodegradable or not biodegradable (inert material) and non-toxic. Both possibilities guarantee the lowest possible impact on the surrounding ecology. The expected impact on the environment should generally be below if the substances are adequately handled and the recommendations of the material safety data sheets are implemented.

11.5 CASE STUDIES ON THE ENVIRONMENTAL AFFECT OF INTOLERABLE TBM ADDITIVES

11.5.1 The Hallandsås Railway Tunnel in Sweden

This tunnel is a classical example of the catastrophic effect of not following the environmental guidelines in tunnel drives (Sposetti 2021). The construction of the tunnel began in 1992 and was originally planned for 1995 to open the service. However, construction was troubled by major difficulties concerning large amounts of water seeping in from surrounding rock, only a small fraction of which had been foreseen. Additionally, the original drill, which was said to drill 100 m per week, broke down after drilling only 18 m since the ground was too soft for a gripper-type TBM. The contractor tried to drill traditionally but had to spend a lot of effort on sealing the water leaks. The contractor went bankrupt and a new contractor took over the job. A scandal broke out when it was learned that a poisonous sealing grout containing 4% acrylamide was used in the grout production. Acrylamide, which is an organic compound with the chemical formula $CH_2 = CHCNH_2$, was used primarily for the production of paper, dyes, and plastics, and in the treatment of drinking water and wastewater, including sewage. Acrylamide is a substance that causes serious effects on human health. The main contractor took no special precautions for the sealant, and he didn't tell to his own workers or to the local people of the risks. By October 1997, local cattle and fish started dying and workers were becoming ill. The local press started an investigation. After tests were done showing high levels of acrylamide contamination, the site was declared a high-risk zone and the sale of agricultural products from the region was banned. By the time construction stopped, a total of approximately 140 t of acrylamide and N-methyl acrylamide were dispersed with the seepage water but had also penetrated into the groundwater. Authorities found out that the chemical product was never properly tested before use in reference to its negative effects on the environment. The information given by the manufacturer and the contractor about the grout was full of contradictions but it was nevertheless clear that the chemical was a poisonous substance, which needed to be handled with care. In addition to killing fish and cattle, a number of people working in the tunnel were intoxicated; fortunately, no one died from exposure to the chemical agents. A total of 29 wells in the area were polluted in this manner. Construction was halted in late 1997. In 2005, construction resumed after a positive decision by the job owner and the Swedish government. The opening ceremony was held on 8 December 2015, and the tunnel opened for traffic on 13 December leaving a tragic and guiding story behind it emphasizing that misusing a chemical in tunnel construction may not only delay the job but it may also increase the construction cost tremendously (Wikipedia 2022).

11.5.2 KISHANGANGA HYDRO ELECTRIC PROJECT HEAD RACE TUNNEL, INDIA

The events occurring in the period of 2011–2014 are summarized by Sposetti (2021) as given below.

The main contractor used the water from a stream next to the tunnel portal as industrial water for TBM works. A concrete water basin was built to capture the water. A very basic water filtering system was built outside the tunnel portal to filter the water before pumping it into the TBM. The return pipeline with water used for mining was discharged indiscriminately into the same stream, a few hundred meters downstream. No one ever measured the concentration levels of chemicals in this water, before and during mining. The cattle of surrounding villages started to die. The contractor neglected the wrongdoing and did not compensate the villagers for damages. Villagers started riots and instability, which remained regularly dismissed by contractors and local authorities. Local Environmental Services (ENV) authorities were absent. Local villagers tried to kidnap foreign EU TBM workers to get attention from the media. Luckily, army intervention, political mediation, and cash compensation were finally used to mitigate the kidnapping risk.

11.6 HYDROCARBON-CONTAMINATED GROUND MANAGEMENT

The hydrocarbon-contaminated ground may be encountered in urban areas like Istanbul where earthquakes frequently occur, loosening the pipe connection of the underground petroleum tanks. This creates two modes for problem drives; one is fire risk during tunnel drives and the second is to find adequate places for storage of the excavated material. This section is devoted to the management of tunnel fire risks during TBM drives in two different metro projects in Istanbul where hydrocarbon-contaminated grounds were met.

11.6.1 Selected Case Stories, Uskudar–Umraniye–Cekmekoy Metro Project

A ground contaminated with petroleum products was encountered in Uskudar–Umraniye–Cekmekoy Metro Project (Gundogdu et al. 2017). The metro line includes 16 stations; tunnels were excavated by both EPB-TBMs and the classical New Austrian Tunneling Method for station tunnels and crosscuts. The water–fuel mixture leakage area consisted mainly of fractured limestone and mudstone–claystone lithology (average overburden is around 30 m). A sudden water inrush occurred along with fuel from an old/unused underground storage tank of a fuel station through one of the forepooling drills and thus construction of the double-tube NATM tunnels with a 30 m^2 cross section was stopped on 12 June 2015. Water–fuel mixture leakage at the tunnel face is seen in Figure 11.2.

The fuel leakage occurred during the excavation of the tunnel face through drilling and blasting, since the face was massive and hard. The amount of leakage of the water–fuel mixture was around 8 L/min in the first 90 min, and then, it dropped eventually to 1 L/min. The contractor cooperated with a team of experts to analyze the situation and determine a plan for restarting the construction activities (Bilgin et al. 2015). It was realized that the water–fuel mixture leakage through the 12 m forepooling pipe was coming from the old/unused (for several years) underground storage tanks of a fuel station. An interview with the administrator of the fuel station indicated that the new fuel tanks were tested recently and there was no leakage in them. It is concluded that the leaking fuel was deposited within the fractures of limestone in the past. The fractured nature of the limestone and the dip and strike of the bedding planes of the mudstone–claystone indicated that the leakage would continue all the way through the ~60 m along the tunnel excavation.

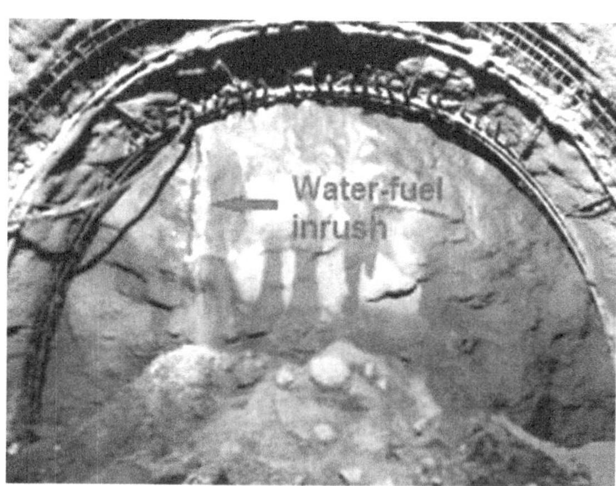

FIGURE 11.2   Water–fuel mixture leakage at the tunnel face. (Bilgin et al. 2015.)

11.6.1.1 Determination of the Safety Risks and Precautions

The most important risk caused by fuel leakage during tunnel excavations is that the tunnel air might become flammable, causing a fire in the tunnel, explosive, suffocating, and poisonous due to evaporating after drainage with water or through the fractures/joints. Therefore, the underground opening should be continuously ventilated with fresh air and the dangerous gasses within the tunnel air should be continuously monitored.

11.6.1.2 Managements of Tunnel Fire Risks During TBM Drive

The following precautions were suggested for the mitigation of the fire risk by Bilgin et al. (2015). Line 1 and 2 tunnels should be continuously ventilated with separate fans. A spare power generator should always be provided and ready on the site. Two gas sensors shall be placed on the roof and invert the tunnel at every 20 m and the measurements shall be transported digitally to the central office in the shaft area. Gas measurements will also be performed with mobile devices. Measurements shall also be performed very close to the face, and the results shall be recorded separately for the top and bottom point measurements. Ten per cent of the lower explosion limit (LEL) at the face and 5% of the LEL at 10 mm behind the face shall be accepted as a "Yellow Alert Level". If the gas amount exceeds the yellow alert level and increases gradually, the alert level shall be increased to "Red Alert Level". All the work in the tunnel should be seized and the tunnel should be evacuated in the red alert condition. Gas measurement personnel should warn all the tunnel crew by sound and light warning systems. An emergency evacuation plan shall be prepared by the representatives of the contractor. All of the machines used in the tunnel should be equipped with fire-distinguishing devices. After an evacuation, the electricity through the tunnel will be cut down and the closest fire department will be informed. The contractor company's crew restarted the construction activities after preparing all of the preventive safety measures on 5 March 2016. They completed the excavation and primary lining jobs on 1 August 2016 without any safety problems.

11.6.2 Ground Contaminated by Gasoline in Bakirkoy–Kirazli Metro Project

Another example of a petroleum product contaminated area is Bakirkoy–Kirazli Mero Line, Istanbul. During site investigations, it was discovered that an area close to the portal of the tubes was contaminated by a leakage of petroleum product coming from a petrol station which is situated near to tunnels. Figure 11.3 shows gasoline found in the ISK-3 borehole located next to the portal of the tubes (Yazici et al. 2019).

As the gas station authorities declared, fuel oil contamination was also discovered under another gas station approximately 60 m north-east of the present gas station. Within a 100 m radius of the working site, there are residential apartments, parks, four schools, and a big shopping mall. During the geotechnical studies, it was observed that the soil around the metro line was permeable. Thus, it was concluded that the leakage around the gas station had contaminated and spread through the surrounding area.

FIGURE 11.3   Gasoline found in the ISK-3 borehole in Bakirkoy–Kirazli Metro Project. (Yazici et al. 2019.)

11.6.2.1   Management Procedure to Cope with Tunnel Fire Risks During EPB-TBM Drive in Gasoline-Contaminated Ground

This example is quite different than the previous example related to the management of tunnel fire risks. According to the National Institute of Occupational Safety and Health, the lower exposure limit to gasoline for a human being is 300 ppm for 8 h, short-term exposure is 500 ppm, and it is easily liable to explosions and fires. For that reason, to keep the hydrocarbon threshold below the exposure limits and explosion limits, several precautions and different management tactics for tunnel fire risk were implemented.

11.6.2.2   Using Bentonite in the Pressure Chamber

First, bentonite was used in order to dilute the concentration of the hydrocarbons within the muck. This application was abandoned after a few attempts since bentonite mixed with other clay minerals found in the excavated material caused clogging of the disc cutter and affected the performance of TBM.

11.6.2.3   Using Gas Sensors Provided by EPB-TBM

The gas sensors installed and integrated into the control system of TBM are $1 \times O_2$, $1 \times CO$, $1 \times CO_2$, $2 \times CH_4$. In case these sensors detect gases above the programmed limit, TBM stops automatically.

11.6.2.4   Establishing a Gas Measuring System Inside of the Pressure Chamber

Gas measurements were taken instantaneously from the excavation chamber through a hose line connected to the bulkhead where methane and hydrocarbon gas sensors were fixed. Gas in the chamber runs through a flow meter and reaches the gas sensor. The flow meter is adjusted to 25 m³/min and the flow rate can be monitored from the operator's cabin. To avoid blockages in the line, an extra water line is connected to the measurement line and in case of blockage, pressurized water is pumped into the line to clean it.

TABLE 11.2
Hydrocarbon Sensors That Are Integrated into TBM

Channel No	Position of Sensors
CH1	Under the screw conveyor discharge gate
CH2	Above the manlock door
CH3	Segment assembly area
CH4	Screw conveyor discharge gate
CH5	Segment feeder
CH6	Hydraulic motors
CH7	Next to the chiller motor
CH8	TBM belt discharge

Source: Yazici, H.A., Okkerman, M.B., Budak, C., 2019. Excavation through contaminated soil with EPB-TBM. WTC 2019, Napoli, Italy, presented by Nuh Bilgin.

11.6.2.5   Establishing New Sensors and a Remote Control System for Gases Generated from Contaminated Ground

To be on the safe side, additional gas sensors that can detect hydrocarbon gases were installed and integrated with the TBM as given in Table 11.2.

According to the hydrocarbon levels, if the gas concentration in the tunnel is between 10% and 20%, sensors give visual and audial warnings (blue light). If the gas concentration is between 20% and 40%, sensors that are integrated with the TBM and PLC stop the excavation and the personnel is evacuated from the tunnel. The tunnel is ventilated until the hazardous gasses are diluted below the threshold limit (orange light). Above 40%, the power is cut off and personnel is evacuated at once (red light). Periodical gas measurements were also carried out with portable gas detectors at various points of the TBM and throughout the tunnel.

11.6.2.6   Ventilation

The airflow was measured at different points of TBM by anemometers. It was detected that there were some areas without airflow reaching. It was presumed that these areas could be some sources for the accumulation of hazardous gases. Due to this reason, the exhaust type of ventilation was abandoned and replaced by blowing type ventilation. Since the contaminated area was close to the shaft, safety engineers decided that any additional ventilation was not required. Exhaust ventilation wasn't chosen since the toxic gas could spread along the tunnel and the air circulation wouldn't be enough to dilute hazardous gas inside the tunnel and the TBM. In this case, if hazardous gases reach unsafe limits, the air velocity would increase.

11.6.2.7   Managing Shift Organization to Control Fire Risk

Excavation works were carried out with the minimum staff possible. Aside from the key crew, no one was allowed inside the tunnel without fire-protective clothes or without the permission of the shift engineer. All activities liable to flame risks were prohibited during TBM advance. Emergency plans were developed and escape roads were kept continuously free. Considering unexpected system defects, calibration errors, or possible damage to gas sensors, manual gas detectors were also used. The excavation in the gasoline-contaminated area was successfully completed. This work was a typical example of how the situation was tackled in such contaminated ground by gasoline. If hazardous gas was detected, the problem was solved with an extra amount of air provided to the tunnel (Yazici et al. 2019).

11.7 METHANE AND H2S-CONTAMINATED GROUND MANAGEMENT

Methane and hydrogen sulfide (H2S) are a potential hazard in tunneling usually resulting in costly delays. These gases may generate an explosive in the tunnel and they may threaten personnel health and project safety when exceeding the allowable limits. As basic rules of management of hazardous gases, accurate gas monitoring systems should be installed inside TBM cutterhead, shield, and segment erector with an automatic TBM shutdown system. Preventive techniques such as grouting, pre-drainage, foam injection, and sealed lining should be used for gassy water inflow (Copur et al. 2012). Gas emissions may increase directly with the water ingress rate, so rapid dewatering systems should be established, as in Zagros Tunnel in Iran (Shahriar et al., 2009).

Methane is a colorless, odorless gas that occurs abundantly in nature and as a product of certain human activities. Methane is lighter than air, having a specific gravity of 0.554. It is soluble in water. Methane in general is very stable, but mixtures of methane and air, with methane content between 5% and 14% by volume, are explosive. Explosions of such mixtures have also been encountered in tunnels, which are explained in detail by Copur et al. (2012). The safety management concerning such cases is summarized in the following chapter titled "Health and Safety in Mechanized Tunneling".

H2S is a colorless gas smelling like rotten eggs. It often occurs naturally in some environments, such as in petroleum fields, gas wells, swamps, and so on. It can also be associated with industrial plants, sewers, sewage treatment plants, or industrial plants. Exposure to low concentrations of H2S may cause irritation to the eyes, nose, or throat. It may also cause difficulty in breathing, headaches, poor memory, and balance problems. Brief exposures to high concentrations of H2S (greater than 1,000 ppm) can cause a loss of consciousness. According to Occupational and Safety and Health Administration (OSHA) regulations, a level of H2S gas at or above 100 ppm is immediately dangerous to life and health. In tunneling, the corrosive effect of this gas may cause big problems, especially for sensitive parts of tunneling equipment. The most comprehensive works and papers on the negative effect of H2S on the performance of mechanized tunneling come from Iran. Shahriar et al. (2009) discussed the Zagros conveyance tunnel in western Iran and highlighted the adverse geological conditions with emphasis on high hydrogen accidents along the tunnel, presenting some practical measures to cope with gassy conditions as well as dealing with potential hydrocarbon contamination and hazardous materials. Mirmehrabi et al. (2012) reported water conveyance tunnel of Aspar driven in H2S-bearing environments. In this paper, they discussed hazards and geological sources of H2S, as well as remedial measures for decreasing the risks and problems in excavation of the tunnel. Morsali and Rezaei (2017) studied the effects of groundwater on H2S and CH4 emissions in Nosoud tunnel in Iran. They concluded that the volume of the released gas was varying with the changes in the groundwater discharge rates and estimation of groundwater inflow into the tunnel was necessary for predicting the volume of gas emissions. However, the latest study released by Su et al. (2020) summarizes the sulfur-bearing formations in the Sichuan Basin in China, whose harmful gases have caused a serious threat to tunnel engineering. They emphasized that tectonic activities have a certain effect on the generation, migration, storage, and sealing of harmful gases, and the influence of fault activities is more obvious.

11.7.1 SELECTED CASE HISTORY ZAGROS CONVEYANCE TUNNEL

Shahriar et al. (2009) discussed the Zagros conveyance tunnel in western Iran and highlighted the adverse geological conditions with emphasis on high emissions of H2S and methane. The Zagros Tunnel project consisted of driving two water conveyance tunnels in western Iran. It is one of the largest tunneling projects undertaken by the Iran Water and Power Resources Development. The first incident of H2S gas detection in the tunnel occurred at the chainage of 3,700 m when the TBM reached the Aspar anticline. In Zagros tunnel, both toxic and explosive gases, including H2S and methane (CH4) have been present simultaneously. The concentration of methane was recorded several times up to 100% of the LEL. According to OSHA regulations, a level of H2S gas at or above 100 ppm is immediately dangerous to life and health. High gas seepage into the tunnel caused serious problems with the tunnel crew and TBM components, while in extreme conditions the recorded H2S of more than 100 ppm caused a four-month shutdown of tunneling operation and work suspension at the tunnel. The following mitigation measures were recommended by Shahriar et al. (2009) to cope with the severity of the situation:

- Installing proper gas monitoring systems inside the TBM cutterhead, shield, segment erector, and several areas in the backup system such as segment crane, track laying area, and control cabin.
- Taking preventive measures for (gas-contaminated) water inflow such as gas drainage and grouting, since the main source of gas emissions into the tunnel is water ingress.
- Using personal protective equipment such as gas masks and capsules.
- Providing sufficient ventilation systems for fresh air to the TBM and face area by using a higher-powered main fan and more efficient booster fans.
- Excavating extra ventilation shafts along the tunnel alignment may be necessary since the efficiency of the ventilation system reduces with the length of the tunnel due to air leakage and related losses along the air ducts.

- Installing rapid dewatering systems by different means.
- Installing gas and fire alarms, and subsequently installing an automatic TBM shutdown system in case of the presence of harmful gases exceeding predetermined levels.
- Using blast-proof equipment and devices to avoid sparking in the electrical system.

11.8 CONCLUDING REMARKS

Management of contaminated ground during and after TBM drives is a major concern in mechanized tunneling. Soil conditioning and using chemicals are inevitable in most tunneling projects. As happened in the past, misusing chemicals creates tremendous environmental problems, destroys nature, causes the death of mammals, etc. like in Hallandsås Tunnel in Sweden and Kishanganga Hydro Electric Project Head Race Tunnel in India. Detailed geotechnical test programs covering petrographical and chemical analysis are necessary prior to the start of the project. Another problem may arise from the excavation, treatment, and storage of the ground already contaminated with carcinogenic chemicals known as PFAS. A typical example of this is the West Gate project in Melbourne. The contractors claim that there is a "Force Majeure Termination Event" since the discovery of contaminated soil was discovered after the contract was signed. There are well-established international standards like OECD guidelines related to the proper selection and use of chemicals in tunneling projects. In the light of a brief summary given earlier, this chapter is intended to emphasize the importance of the management of contaminated ground during and after TBM drives.

REFERENCES

Bilgin, N., Balci, C., Aslanbas, A., 2021. Case studies leading to the management of tunnel fire risks during TBM drives in an old coalfield. *Tunnelling and Underground Space Technology,* 112, p. 103902.

Bilgin, N., Copur, H., Fisne, A., 2015. *Analysis of Fuel Leakage and Precautions for Restarting the Construction Works. Report Submitted to Dogus Construction Inc.,* Istanbul Technical University Project (in Turkish).

Copur, H., Çınar, M., Ökten, G., Bilgin, N., 2012 A case study on the methane explosion in the excavation chamber of an EPB-TBM and lessons learnt including some recent accidents. *Tunnel Underground Space Technology,* 27(1), pp. 159–167.

Edwards, J., 2021. West Gate tunnel project blows out by \$3.3 billion as transurban scraps estimated completion date, *News, Posted* Monday 9 August 2021.

EFNARC, 2005. *Specification and Guidelines for the Use of Specialist Products for Mechanized Tunnelling (TBM) in Soft Ground and Hard Rock.* European Federation Dedicated to Specialist Construction Chemicals and Concrete Systems.

Gundogdu, I., Kalayci, G., Isler, M.K., Bilgin, N., Copur, H., Fisne, A., 2017. Safety precautions against fuel leakage into NATM tunnels of Uskudar–Umraniye–Cekmekoy Metro Project, in *ITA World Tunnel Congress (WTC 2017), Surface Challenges – Underground Solutions.* Bergen, Norway.

Langmaack, L., 2017. How to protect our nature – Chemistry in TBM tunnelling. from the laboratory to on-site use and muck disposal, in *Proceedings of the World Tunnel Congress 2017 – Surface Challenges – Underground Solutions*. Bergen, Norway.

Mirmehrabi, H., Ghafoori, M., Lashkaripour, G., Azali, S.T., Hassanpour, J., 2012. Hazards of mechanized tunnel excavation in H2S bearing ground in Aspar tunnel, Iran. *Environmental Earth Sciences*, 66(2012), pp. 529–535, doi:10.1007/s12665-011-1262-y

Morsali, M., Rezaei, M., 2017. Assessment of H2S emission hazards into tunnels: the Nosoud tunnel case study from Iran. *Environmental Earth Sciences* 76(2017), p. 227. doi:10.1007/s12665-017-6493-0

OECD 203, 2000. Acute toxicity testing, Organisation for economic co-operation and development, Paris 2000. *ENV/JMMONO*, p. 6.

OECD 301, Biodegradability testing, Organisation for economic co-operation and development, Paris 1995. *OECD/GD* 95, p. 43.

OECD, 2005. Guideline for testing of chemicals. *ENV/JM/TG* 5(1), April 2005.

Peila, D., Picchio, P., 2011. Influence of chemical additives used in EPB tunneling and its management. *Materials Science, Corpus ID: 134538329*, pp. 50–95.

Rowland, J., 2020. Contaminated soil delays Melbourne West Gate TBMs, *Tunnel Talk* 23 April.

Shahriar, K., Rostami, J., Khademi, J., 2009. TBM tunneling and analysis of high gas emission accident in Zagros long tunnel, in *Proceedings of the ITA-AITES World Tunnel Congress*, Budapest, Hungary.

Sposetti, M., 2021. Environmentally friendly solutions for the TBM tunnelling industry, *Tunnel Turkey 2021, International Tunneling Symposium*, 21–22 December, Istanbul.

Stewart, R., 2019. *Environmental Science in the 21st Century. Texas A&M University*, Open (Access) Textbook https://web.archive.org/web/20160310085142/http:// oceanworld. tamu.edu/resources/environment-book/groundwaterremediation.html.

Su, P., Zhao, Y., Xu, Z., Du, Y., Qiu, P., Wang, D., Li, Y., 2020. Study on the distribution of sulfur-bearing formations in the Sichuan Basin and its damage to tunnel engineering, 17th World Conference ACUUS 2020 Helsinki, IOP Conf. *Series: Earth and Environmental Science* 703(2021), p. 012030, doi:10.1088/1755-1315/703/1/012030

Vinai, R., Oggeri, C., Peila, D., 2008. Soil conditioning of sand for EPB applications: A laboratory research. *Tunnelling and Underground Space Technology*, 23, pp. 308–317.

Wikipadia, 2022. https://en.wikipedia.org/wiki/Hallands%C3%A5s_Tunnel.

Willingham, R., 2020. West gate tunnel builders seek to terminate contract over contaminated soil, Transurban tells ASX, *News, posted* Wed 29 Jan 2020. www.tunnelsonline.info/ news/australias-largest-tbm-starts-on-west-gate-tunnel-9524299, Downloaded on 13 October 2022.

Yazici, H.A., Okkerman, M.B., Budak, C., 2019. Excavation through contaminated soil with EPB-TBM. *WTC 2019*, Napoli, Italy.

12 Risk Management in Mechanized Tunneling Projects and Insurance Aboutissement

12.1 INTRODUCTION

Risk management is one of the main issues in the success and efficiency of tunneling projects. Significant delays in project scheduling, cost overruns, and increases in hazards in major projects show that risk management is not being given the importance it should have. In order to be able to measure and control current costs and the schedule against set targets, it is necessary to assess risks in a transparent manner and to consider them appropriately. Risk management is the systematic process of identifying, analyzing, evaluating, and monitoring hazards and associated risks, and also allocating risks to the various parties to the contract. The risk is a measurable part of uncertainty, for which we are able to estimate the occurrence probability and the size of the damage. The risk is assumed as a deviation from the desired level and is regarded as the analysis of adverse events even at the stage of planning and programming a construction project.

A financier or a bank creditor when making his assessment relies on whether the client has taken a risk-based approach in choosing the tunnel alignment depending on a reliable baseline report. He will also check whether the client has established a risk management framework based on quality, and cost or he has used risk registers and got design checkers. The decisions of financiers and insurers are linked to each other. The insurers assess the project and if it aligns with the code of practice for risk management of tunnel works (ITIG 2016), and if they determine that risk management is solid, they give the decision of insurability, since the development of the code of practice is an important step in building confidence in tunneling projects (Ballantyne 2018). Due to the interrelationships between risk management, code of practice, and insurance issues, these three parameters will be handled together in this chapter, giving some examples from different tunneling projects.

12.2 RISK IDENTIFICATION, EXAMPLES OF RISK ASSESSMENT

In the design stage before starting a tunneling project, all risks should be identified as possible and should be assessed. Geological uncertainties sometimes make this process a difficult task. In such cases, interpretation of geologic reports based on past experiences will help the decision-maker to a great extent. First, this section will be

DOI: 10.1201/9781003358978-12

devoted to risk classification for tunnels to be excavated close to major fault zones, leading to the decision on the excavation method, tunneling with the conventional method, or tunneling with TBM.

12.2.1 TBM Risk Classification for Tunnels To Be Excavated Close to Major Fault Zones

Nurdağı high-speed railway tunnel consists of two tubes. Each tube has a length of 9,750 m. The excavation is planned to start from chainage 13 + 450 m and terminate in chainage 3 + 700 m. The chainage from 13 + 450 to 12 + 400 m consists of the Karadağ Limestone of the Mesozoic age, which is affected by the East Anatolian Fault (EAF), fracturing the rock formation to a great extent. Turkey is in a tectonically active region that experiences frequent destructive earthquakes. On a large scale, the tectonics of the region is controlled by the collision of the Arabian Plate and the Eurasian Plate. At a more detailed level, the tectonics become quite complicated. A large piece of continental crust almost the size of Turkey, called the Anatolian block, is being squeezed to the west. The block is bounded to the north by the North Anatolian Fault and to the southeast by the EAF. The EAF is a major strike-slip fault zone in eastern Turkey. It forms the transform-type tectonic boundary between the Anatolian Plate and the northward-moving Arabian Plate. High water ingress is expected in this area. Karadağ Limestone discharges the water at the toe of the mountain on the Nurdağı side. Several springs are available along EAF. After the risk analysis which is explained later, due to technical difficulties and the time necessary to provide TBM, the first 1,050 m in limestone was opened by the traditional tunneling method, i.e., drill and blast. The geologic cross section of this area is planned to be opened by the drill and blast method, as seen in Figure 12.1.

The geological formation from the chainage 12 + 400 to 4 + 850 m consists of the middle Ordovician aged Kızlaç Formation of very massive interbedded meta-sandstone, meta-quartzite, and meta-mudstone having very high strength and abrasive characteristics. This section is planned to be excavated with a single shield hard

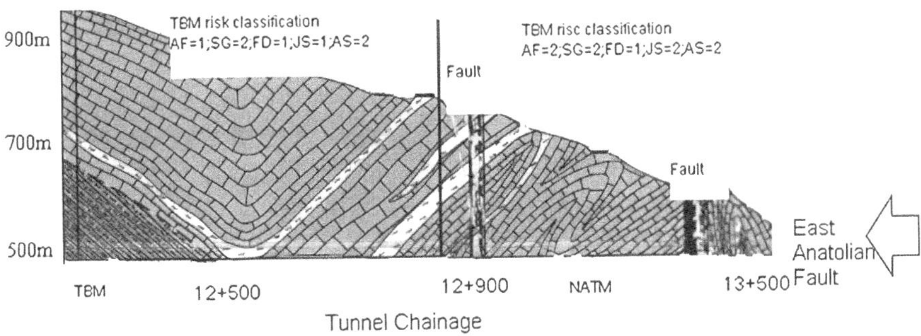

FIGURE 12.1 Geological cross-section of the tunnel Nurdağı tunnel between chainage 13 + 450 to 12 + 400, with TBM risk classification given in Table 12.1. (Bilgin, 2016.)

TABLE 12.1

A TBM Risk Classification for Tunnels to Be Excavated Close to North and South Anatolian Faults

Factors Effecting the Risk of Using TBM	Classification
(1) Distance of the tunnel to NAF and EAF. The possibility of tectonic stresses (AF	(1) Within 0.5–2 km to NAF and EAF; (2) Very close to NAF and EA
(2) The possibility of high amount of water ingress into the tunnel. Detailed geological reports and careful observation of drilling logs are necessary (SG)	(1) Less than 100 L/s (2) More than 100 L/s
(3) The possibility of seeing geological discontinuities in front of tunnel face. The criterion is that in NATM it is easy to see and control geological discontinuities in the tunnel face (FD)	(1) For NATM excavation method (2) For TBM excavation method
(4) Geological discontinuities (JS)	(1) $Q < 1$ RMR < 20; (2) $Q > 1.1$ RMR > 21
(5) The presence of anticlinal and synclinal (AS)	(1) One within 1 km; (2) More than one within 1 km.

Source: Bilgin (2016).

Note: If the total mark is between 8 and 10, it is very risky to use TBM; if the total mark is between 5 and 8, it is risky to use TBM; if the total mark is between 2 and 5, the risk of using TBM is in medium level; if the total mark is 0–2, using TBM is not risky.

rock TBM. However, the massive character of the rock formation could change from 4 + 850 to 3 + 700 m, being affected by local faults and shear zones. In this zone, RQD values are very low, and high water ingress is also expected. This section is planned to be excavated with NATM tunneling method. The tunnels excavated close to North and EAFs led to the development of a risk classification method defined in Table 12.1. According to this table, the use of TBM in the Nurdağı tunnel between 13 + 500 and 12 + 800 m is very risky, 12 + 800 to 12 + 500 m is risky, and favorable up to 4 + 850 m.

12.2.2   RISK CLASSIFICATION FOR METHANE MIGRATION FOR TBM EXCAVATING UNDER AN OLD COAL FIELD

The new third Istanbul Airport has been constructed 35 km away from the city center. It is in the northwest of the city center towards the Black Sea coast, on the European side of Istanbul. Side investigations revealed the fact that a risk of methane and coal dust explosion would be a major concern during TBM drives due to an old coalfield area. The past experiences gained from coal mines and tunnel accidents guided the authors to prepare emergency plans to manage methane explosions and tunnel fires during TBM drives. The possibility of the occurrence of methane in the excavated area was a big concern, since, in the past, a methane explosion occurred inside the excavation chamber of Selimpasa tunnel in Istanbul, 80 km away from the current project (Copur et al. 2012). Safety and health standards of the Occupational Safety and Health Administration (OSHA) of the USA require continuous monitoring of gases for underground construction when TBMs are used. OSHA states that when an air sample indicates 5% LEL or more, air supply should be increased with continuous gas control in order to decrease the accumulation of methane; at 10% LEL or more, hot works such as welding or cutting should be suspended, since the heat generated from hot works is at the level to make an explosion; at 20% LEL or more, power should be turned off and crew should be withdrawn (Kissel 2006). Methane accumulates at the crown of a tunnel, so it is strictly essential to make methane measurements at the crown. If the methane content passes a predetermined value, TBM is automatically shut down. However, another important property of methane is that it dissolves in groundwater and travels kilometers away from its source of emission and gets free where the pressure decreases, especially in the water in the form of bubbles. So it is also necessary to use portable methane measuring devices apart from sensors established within the TBM in proper places.

The following security program for management of methane/dust explosion and fire risk was initiated within the chainages where the coal seams were excavated: (a) smoking is strictly prohibited in the tunnel; (b) never forget that the methane content in the pressure chamber of EPB-TBM is always higher than the methane content in the tunnel, since the methane in the air outside of the tunnel is always diluted by tunnel ventilation and the oxygen in the foam is a very favorable medium for methane explosion; (c) obey strictly the standards given by OSHA; (d) never try a high daily advance rate, since methane emission per m^3 of excavated coal is constant and higher advance rate means higher methane content in the tunnel air, so a mean daily advance rate of 22.7 m was reached within the change studied, although a daily advance rate more than 40 m/day was possible to obtain; and (e) avoid the accumulation of coal dust around the belt conveyor and TBM area, and don't forget that coal dust explosion is most dangerous than methane explosion (Aslanbas and Bilgin 2020). The geology is very complex along the first metro line. Quarterly, Tertiary, and Paleozoic Formations are found in the area and there are several transition zones along tunnel routes, creating several problems during tunnel excavation. The airport construction area was an old coal exploded area. Oligocene-aged coal seams consisted of the upper seam, middle seam, and lower seam. Thin clay bands existing within the coal seams decreased the calorific value of the end product. The mine is currently abandoned and one of the biggest airports in the world is being constructed in this area. The thickness of the upper coal seam is usually greater than 1.5 m, and the lower coal seam has a thickness of 0.5 m, and they are usually separated by clay bands as shown in Figure 12.2.

In most cases, the middle coal seam is missing from the geological sequence. Cleats are more dominant in the upper coal seam, having the capability of having more methane gas than the other coal seams. The coal was found to be liable to dust generation during the excavation. During the excavation, variation of methane

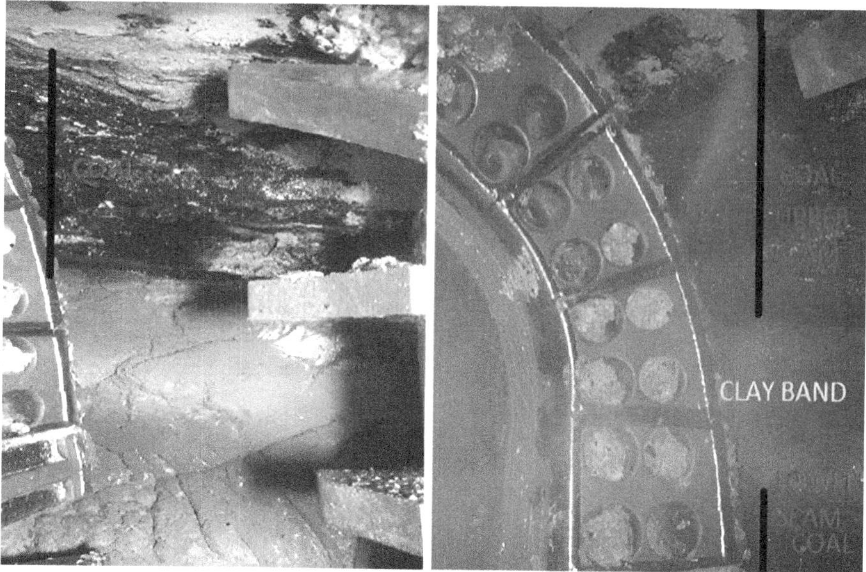

FIGURE 12.2 View of coal seams through of cutterhead openings. (Aslanbas and Bilgin 2020.)

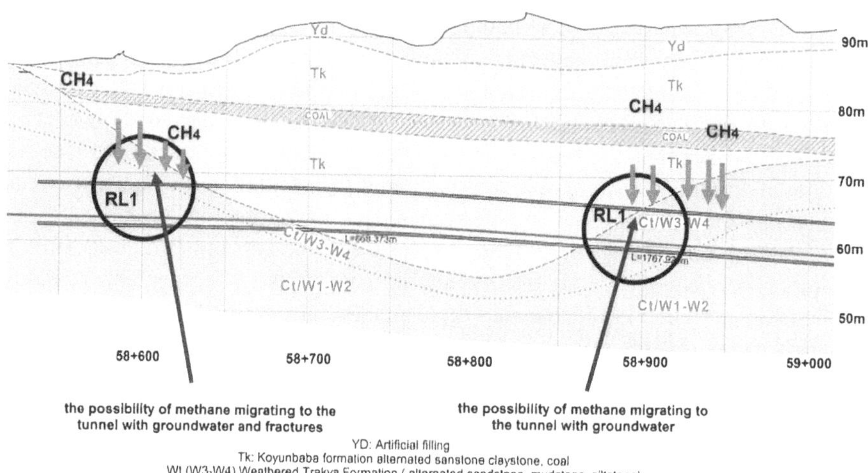

FIGURE 12.3 Geological cross-sectional areas with a possibility of methane migration at risk RL1. (Aslanbas and Bilgin 2020.).

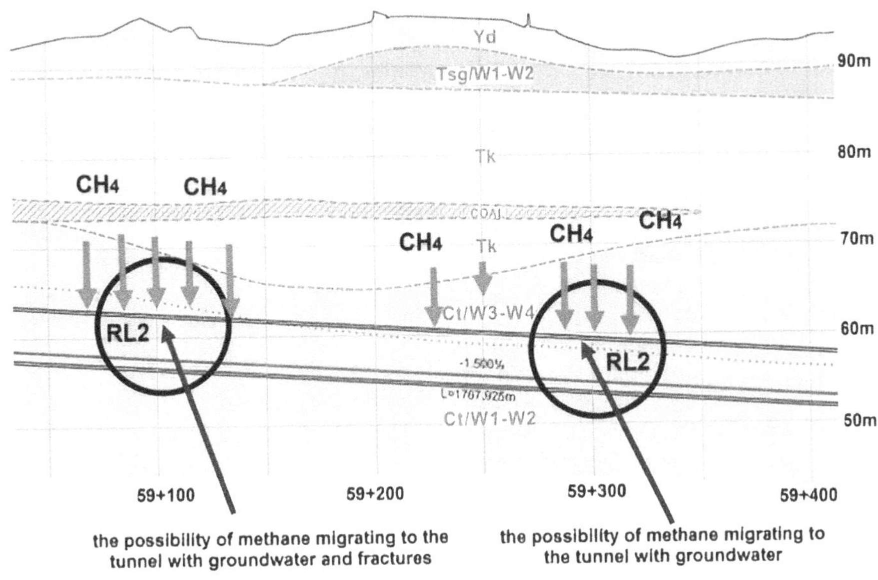

FIGURE 12.4 Geological cross-sectional areas with a possibility of methane migration at risk RL2. (Aslanbas and Bilgin 2020.)

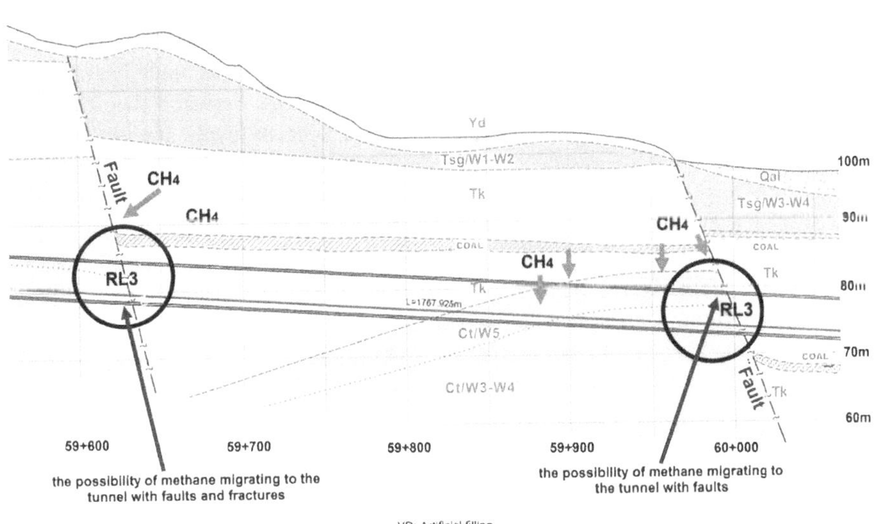

FIGURE 12.5 Geological cross-sectional areas with the possibility of methane migration at risk RL3. (Aslanbas and Bilgin 2020.)

TABLE 12.2
Classification of Risk Levels for Methane Migration to the Tunnel

Risk Level	Geologic Structure, Rock Discontinuities, Fractures	Distance Between Coal Seam and the Crown of the Tunnel	Distance of the Ground Water Level to the Coal Seam
RL3, Very high	Distance to fault up to 5 m	3–5 m	3–13 m
RL2, High	At the edge of the anticlinal, within the transition zone and fractured rock	10 m	5 m
RL1, Moderate	The formation between coal seam and tunnel is fractured	10–15m	7–8 m

Source: Bilgin et al. (2021).

content in % LEL was found to change between 3.1% and 2.6% (Aslanbas and Bilgin 2020).

The two tubes of the second line of Istanbul New Airport pass under a coal seam having a thickness of 1–3 m between 58 + 600 and 60 + 000 km as seen in Figures 12.3, 12.4, and 12.5. The coal seam is close to the tunnel from a few meters to 10 m and is reported to have methane. It is classified as gassy and risky for TBM drives at three levels: RL1, RL2, and RL3. This classification is made based on the geological structure, groundwater level, and the distance between the coal seam and the crown of the tunnel as given in Table 12.2. Risk levels are illustrated in Figures 12.3–12.5

Measures taken based on risk level are basically the same as mentioned earlier. However, a training course for the tunnel staff and workers is recommended on methane risk levels. At risk levels RL1, RL2, and RL3, the amount of air provided to the tunnel should be increased by an amount of 50%. However, at risk level RL3, the continuous communication between TBM crew, shift engineer, and safety engineer should be at the maximum level at least once per hour.

12.3 ANALYZING RISKS, EVALUATING RISKS, MITIGATION THROUGH ENGINEERING SOLUTIONS

A risk analysis carried out on Üsküdar–Ümraniye–Çekmeköy–Sancaktepe, M5 Metro line, will be used as an example. It is 16.9 km in length and has 16 stations (Figure 12.6). The construction of the project was started in March 2012, and the line was opened on 15 December 2017. An average daily advance rate of 18 m was a prerequisite of the project. During the sinking of the station shafts, the contractor realized that there were several geological uncertainties along the metro line.

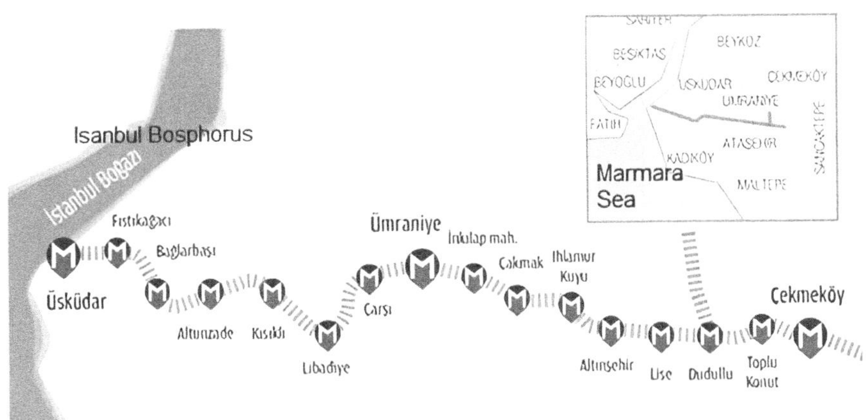

FIGURE 12.6 Üsküdar–Ümraniye–Çekmeköy–Sancaktepe, M5 Metero line with 16 stations.

Therefore, he asked the university to review the geology and carry out a TBM performance analysis to estimate daily advance rates, including a risk analysis aimed at scheduled job termination. In this section, the research study carried out for this project will be summarized in order to give an illuminating idea of how to carry out a risk analysis step by step (Bilgin et al. 2013). The readers are encouraged also to overview the paper titled "Geotechnical risk assessment-based approach for rock TBM selection in difficult ground conditions" by Shahriar et al. (2008), and the paper titled "Application of a methodology for risk management on tunnel project" by Gaillard et al. (2013).

12.3.1 Revision of Geological and Geotechnical Report

In addition to the previous studies, extra core drilling and geomechanical tests were carried out to make the geological profile more comprehensive. Detailed in-situ observations were made in Ihlamur, İnkilap, Çarşı, Altunizade, Libadiye, and Kısıklı shafts, including in situ N-type Schmidt hammer tests. A typical cross section and general view of the tunnel face is given in Figures 12.7 and 12.8, and some of the geotechnical test results from the new study are given in Table 12.3.

12.3.2 Prediction of the Mean Daily Advances

A model to predict daily advance rates of EPB-TBMs in complex geology in Istanbul was used in the risk analysis. This method, which was recently described by Namlı et al. (2013 and 2014) and Namlı and Bilgin (2017), depends on the prediction of field-specific energy, torque consumed by TBM, and machine utilization time. In the model, the effect of EPB face pressure is also considered. The calculation made showed that it was not possible to achieve the average prerequisite value of 18 m daily advance rates. The model predicted that the average daily advance rate could only reach 8.5–11.2 m, including all delays. However, in the

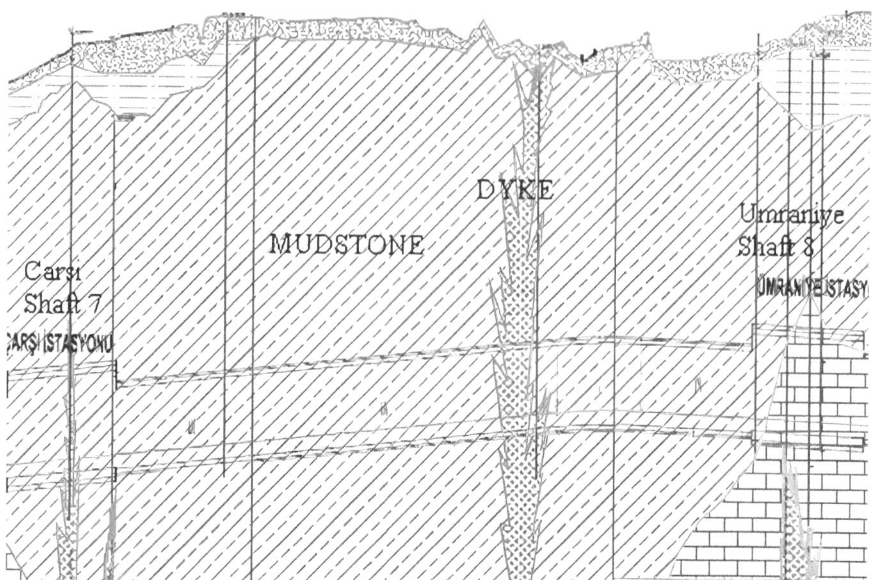

FIGURE 12.7 The geological profile between Çarşı and Ümraniye Station. (Bilgin et al. 2013.)

FIGURE 12.8 Tunnel face of Çarşı Metro Station (shaft 7). (Bilgin et al. 2013.)Note: Andesite is seen on the left, and mudstone on the right.

TABLE 12.3
Some of the Geotechnical Test Results Obtained in the New Study

The Shafts Where Measurements Were Taken	Formation	In situ Schmidt Hammer N-type Massive Formation (+/−ss)	In situ Schmidt Hammer N-type Fractured Formation (+/−ss)	Uniaxial Strength (MPa +/−ss)	Cerchar Abrasivity
Ihlamur Shaft	Arkose-sandstone	44 +/− 3	34.3 +/− 3	30 +/− 2	3
İnkilap Shaft	Limestone	52 +/− 4	36.4 +/− 5	90 +/− 3	--
Çarşı Shaft	Andesite	57 +/− 5	---	117 +/− 12	5
Çarşı Shaft	Mudstone	49 +/− 1	40 +/− 3	98 +/− 3	-
Altunizade Shaft	Andesite	45 +/− 8	27 +/− 2	86 +/− 17	4.5
Altunizade Shaft	Mudstone	45 +/− 4	33 +/− 4	54 +/− 2	--
Libadiye Shaft	Limestone	64 +/−2	52 +/− 2	120 +/− 5	–
Ümraniye Shaft	Limestone	64 +/ − 2	52 +/− 2	139 +/− 5	–
Kısıklı Shaft	Quartzite	58 +− 1	–	–	5.5
Kısıklı	Conglomerate	–	–	–	3.5

Source: Bilgin et al. (2013).

report provided to the contractor, it is strictly emphasized that these numbers were only valid if identified risk and mitigation measures pointed were considered with great care.

12.3.3 Risk Identification and Mitigation Measures

The process of risk identification may rely upon a review of worldwide operational experience of similar projects written, upon the study of generic guidance on hazards associated with the type of work being undertaken, and upon discussions with qualified and experienced staff from the project team and other organizations around the world (Eskesen et al. 2004). Based on the past experiences obtained in the tunnels excavated in the complex geology of Istanbul, identification risks and mitigation measures for Üsküdar–Ümraniye–Çekmeköy–Sancaktepe metro project were developed as given in Table 12.4. Mitigation measures should be followed carefully in order to reach the targets of risk analysis.

Risk rating system frequency assignments (F) is made as given in Table 12.5.
Risk rating system consequence assignments (C) is given in Table 12.6.
Risk score is calculated as given below.

Risk score = Frequency (F) × Consequence (C)

The effect of risk score on the impact of the event is given in Table 12.7

TABLE 12.4
Identification Risks and Mitigation Measures for Üsküdar–Ümraniye–Çekmeköy–Sancaktepe Metro Project

Identified Risks	Mitigation Measures
1. High level of surface settlements due to small overburden values	The control of excavated volume and compared with theoretical volume, apply the proper EPB pressure, annulus grouting should be done properly, frequent control of wire brushes
2. Ancient water wells may cause excessive surface settlements as reported in Güclücan et al. (2008)	Detection of ancient water wells and filling them with concrete
3. Water ingress from sea in Üsküdar area	Ground freezing or chemical injection
4. Dynamic loads coming to the main bearing in transition zones between the main rock formation and dykes	Working with low rotational speed (rpm) and frequent analysis of the oil taken from the main bearing
5. Dynamic loads due to transition zones between the main rock and dykes causing disc cutter failures as reported in Güclücan et al. (2008)	Working with low rotational speed (rpm)
6. Squeezing or blocking the cutterhead in transition zones.	The excavated volume should not exceed the theoretical volume, frequent control of grizzly bars, the ratio of the thrust to torque is good indicator of squeezing or blocking the cutterhead. It should be watched carefully
7. In transition zones between the main rock and dykes, big block coming to the face may damage the cutters.	Frequent control of grizzly bars
8. Changing the direction of the TBM in the interface between hard and soft formations	Working with low thrust and rotational speed
9. Abrasivity of the rocks which may cause excessive disc consumption	Working with proper foams, especially anti wear agent foams
10. Decreasing the penetration and increasing specific energy causing low penetration rates in very high strength rocks as experienced in Beykoz tunnel (Güçlücan et al. 2008)	Working with V type disc cutters
11. Flow heaves in high EPB pressures	Working with proper EPB face pressure

Source: Bilgin et al. (2013).

TABLE 12.5
Risk Rating System Frequency Assignments (F)

Rating (F)	Occurrence Interval
1. Rare	May occur only in exceptional circumstances
2. Unlikely	Event is unlikely to occur, but it is possible during the execution of the project
3. Possible	Event could occur during the period of the execution of the project
4. Likely	Event likely occurs once or more during the execution of the project
5. Almost certain	Event likely occurs many times during the execution of the project

Source: Bilgin et al. (2013).

TABLE 12.6
Risk Rating System Consequence Assignments (C)

Probability (C)	Impact on Daily Advance
1. Insignificant	Delays less than 1 day
2. Minor	Delays around 1–2 days
3. Moderate	Delays around 2–7 days
4. Major	Delays around 7–30 days
5. Catastrophic	Delays more than 30 days

Source: Bilgin et al. (2013).

TABLE 12.7
The Effect of Risk Score on the Impact of the Event

Risk Score	Impact of the Event
0–8	Tolerable
9–15	ALARP – as low as reasonably possible
16–25	Non-tolerable

Source: Bilgin et al. (2013).

To give examples to readers, risk scores between Ümraniye–Çarşı and Fıstıkağacı–Üsküdar shafts are given in Tables 12.8 and 12.9.

After submitting the risk analysis, the contractor Doğuş, and the job owner Istanbul Municipality, changed the prerequisite daily advance from 28 m 8.5–11.2 m, and the project was successfully terminated in schedule time.

TABLE 12.8
Risk Scores Between Ümraniye and Çarşı Shafts (791.6 m in Length)

Identified Risk	Frequency (F)	Consequence (C)	The Score of Risk
1	3 Possible	2 Delays around 1–2 days	6 Tolerable
2	2 Unlikely	3 Delays around 2–7 days	6 Tolerable
3	–	–	–
4	1 Rare	2 Delays around 1–2 days	2 Tolerable
5	3 Possible	3 Delays around 2–7 days	9 ALARP
6	4 Likely	3 Delays around 2–7 days	12 ALARP
7	4 Likely	3 Delays around 2–7 days	12 ALARP
8	4 Likely	3 Delays around 2–7 days	12 ALARP
9	2 Unlikely	2 Delays around 1–2 days	4 Tolerable
10	2 Unlikely	2 Delays around 1–2 days	4 Tolerable
11	2 Unlikely	2 Delays around 1–2 days	4 Tolerable

Source: Bilgin et al. (2013).

TABLE 12.9
Risk Scores Between Fıstıkağacı Üsküdar Shafts (1043.3 m in Length)

Identified Risks	Frequency (F)	Consequence (C)	The Score of the Risk
1	3 Possible	4 Delays around 7–30 days	12 ALARP*
2	3 Possible	4 Delays around 7–30 days	12 ALARP
3	5 Almost certain	5 Delays more than 30 days	25 Non-tolerable
4	3 Possible	3 Delays around 2–7 days	9 ALARP
5	5 Almost certain	3 Delays around 2–7 days	15 ALARP
6	5 Almost certain	3 Delays around 2–7 days	15 ALARP
7	5 Almost certain	3 Delays around 2–7 days	15 ALARP
8	2 Unlikely	1 Delay less than 1 day	2 Tolerable
9	3 Possible	3 Delays around 2–7 days	9 ALARP
10	2 Unlikely	2 Delays around 1–2 days	4 Tolerable
11	3 Possible	3 Delays around 2–7 days	9 ALARP

Source: Bilgin et al. (2013).

Note: ALARP – as low as reasonably possible.

12.4 RISK MONITORING AND REVIEWING, CODE OF PRACTICE FOR RISK MANAGEMENT OF TUNNEL WORKS

Artopoulos (2016) analyzed 29 known tunnel losses and reported that the costs for the studied incidents are approximated to the overall amount of US$800 million. He also reported that the vast majority of these losses were attributed to tunnel collapses (nearly 83%), with the remaining ones being due to flood (10%) and fire (7%). The most frequent causes of these losses were (a) insufficient ground investigation or interpretation, (b) faulty design or workmanship, and (c) lack of appropriate measures or procedures in place that would enable the timely recognition of forthcoming hazards and hence the implementation of the appropriate corrective actions (Artopoulos 2016).

After the collapses, flooding and fires in several tunnels before 2000, thinking that tunnel projects were becoming uninsurable, the Association of British Insurers, and the British Tunnelling Society worked together to create a new benchmark for best practice. Published in 2003, the "Joint Code of Practice for Risk Management of Tunnel Works in the UK" was followed by an international version in 2006 named the "Code of Practice for Risk Management of Tunnel Works". These codes became known worldwide as the Tunnel Code of Practice or TCOP (Ballantyne 2018). However, they are not legally binding but they do contain some important principles which aim to make all parties involved to take the right approach to risk identification, control, and elimination.

Big insured accidents make tunnel construction more expensive. The Codes, therefore, aim to unify and determine a minimum standard of risk control methodology, in other words they were developed as a professional risk management tools. The Codes cover all phases of underground construction activities, i.e., the preparations, engineering, project allocation, and implementation. The Codes emphasize the insurers' involvement in the contract. The following are among the main principles of the Codes (ITIG 2016) which are summarized in an article titled "Risk in Underground Construction" as given below (Law Explorer 2015):

- *The requirement to submit the Register of Risks. The Register of Risks is an open document (it is possible and desirable to extend it during the course of construction), which clearly defines to whom a risk belongs, how it is to be controlled and how it is to be mitigated. The Register of Risks is a part of a Quality Control System, being, as such, subject to independent audits.*
- *Use of the "standard forms of contract" and technical standards.*
- *The contract should include a risk allocation and sharing clause, concerning geology or unforeseeable physical conditions.*
- *The contract should include a provision regulating the geo-monitoring process.*
- *The contract should include a provision allowing variations and implementing value engineering.*

- *The employer must have sufficient knowledge of geological risk control. If lacking this knowledge on the side of its own staff, they are obliged to hire a consultant or a contractor able to meet this requirement.*
- *The employer is obliged to invest sufficient funds in geological and hydro-geological surveying. This allows bidders to prepare and price the offer in respect of the known ground conditions and related risks in connection with tunnelling.*
- *The employer is also obliged to have sufficient funds and time for project preparation.*

A project may be deemed "uninsurable" where the above requirements are not met. This, in itself, is a project commencement obstacle.

A flowchart indicating the most essential issues/tasks needing to be accurately fulfilled in the frame of the risk assessment and risk management processes during the design and construction stages of tunnel works to ensure compliance with the code of practice is given in Figure 12.9 (Artopoulos 2016). It is important to note that the most neglected part of risk management is the appointment of risk management team tasks. The insurance inspectors usually notice that the documents missing in the claim of the contractors are documentation on regular or extraordinary meetings (progress reviews), decision-making, and follow-up documents as issues of MoM.

12.5 INSURANCE AND CLAIM ISSUES

Compared to many other construction projects carried out on the ground surface, identification and evaluation of risks in tunneling are the most challenging ones in the wide spectrum of engineering insurance due to the great number of uncertainties inherent in the underground, demanding extensive side investigation for risk reduction and management. In tunneling construction, unexpected ground conditions, faulty design, and bad workmanship are the main causes of delays, cost overruns, and surely insurance payouts.

In spite of the huge experience and technological advancements accumulated in the tunnel construction industry, tunneling insurance has been proven to be a risky business as revealed by the great frequency of major tunnel losses as reported in Table 12.10. This table is compiled from Artopoulos's (2015) and Wannick's (2016) works. However, the causes of losses are only mentioned in Artopoulos's work: "More especially, the increased number of costly tunnel construction disasters since the early '90s started creating serious 'headaches' for the insurance community, as they threatened to reverse in the medium terms the insurance of tunneling risks to unprofitable business" (Artopoulos 2015).

One interesting thing is that geological uncertainties are not mentioned in Table 12.10 as one of the causes of tunnel failures. However, the headrace tunnel at the Glendoe Hydro Electric Plant in Scotland is probably one of the examples to be mentioned in tunneling history in this respect. The water-bearing headrace tunnel was constructed through hard rock using a TBM and was designed to be substantially

1. Risk Assessment

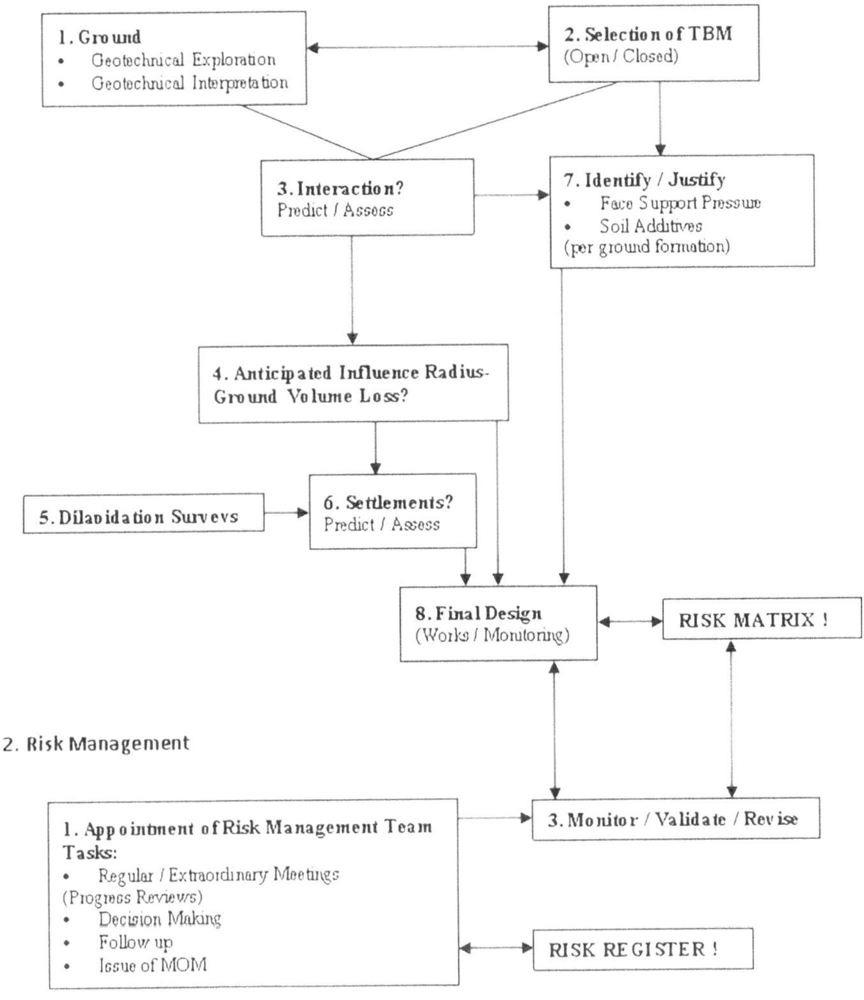

2. Risk Management

FIGURE 12.9 The most essential issues/tasks needing to be fulfilled for the risk assessment and risk management processes during tunneling as to ensure compliance with code of practice. (Artopoulos 2016.)

unlined. In August 2009, just months after it had opened Glendoe the tunnel collapsed, blocking a 71-m length, which necessitated the construction of a by-pass tunnel (Hencher 2019). The tunneling contract was design-build. A dispute arose between the client and the contractor concerning liability for the collapse, and who should bear the cost of the remedial works as they proceeded. The client went to court claiming that the collapse and secondary features were due to defective design because the

TABLE 12.10

Insurance Payouts of Some Major Tunnel Failures from 1994 to 2005

Date	Project and Country	Event	Cause	US$ (millions)
1994	Great Belt Link, Denmark	Fire	Faulty design?	33
1994	Munich Metro, Germany	Collapse	Faulty design	4
1994	Heathrow Express Link, GB	Collapse	Bad workmanship	141
1994	Metro Taipei, Taiwan	Collapse	Bad workmanship	12
1995	Metro Los Angeles, USA	Collapse	Bad workmanship	9
1995	Metro Taipei, Taiwan	Collapse	Bad workmanship	29
1999	Hull Yorkshire Tunnel, UK	Collapse	Faulty design	55
1999	TAV Bologna – Florence, Italy	Collapse	Faulty design/ Workmanship	9
1999	Anatolia Motorway, Turkey	Earthquake	Earthquake/Faulty design	115
2000	Metro Taegu, Korea	Collapse	Faulty design	24
2000	TAV Bologna – Florence, Italy	Collapse	Faulty design/ Workmanship	12
2002	Taiwan High-Speed Railway	Collapse	?	30
2002	SOCATOP Paris, France	Fire	?	8
2003	Shanghai Metro, PRC	Collapse	Faulty design	80
2004	Singapore Metro, Singapore	Collapse	Bad workmanship	t.b.a
2005	Barcelona Metro, Spain	Collapse	Bad workmanship	t.b.a

Source: Artopoulos (2015) and Wannick (2016).

Note: The total for these 19 major losses was estimated at about US$600 million.
t.b.a used to say that some details of an event have not yet been decided.

rock support which the contractor had installed in the tunnels was inadequate. The client claimed £200 million from the contractor, including £130 million for the cost of repairs, and £65 million for loss of profit. The clause in the question of the New Engineering Contract (NEC) provided that "the Contractor is not liable for Defects in the works due to his design so far as he proves that he used reasonable skill and care to ensure that his design complied with the Works Information" (Doran 2017). The NEC, or NEC Engineering and Construction Contract, is the system created by the UK Institution of Civil Engineers that guides the drafting of documents on civil engineering, construction, and maintenance projects for the purpose of obtaining tenders, awarding, and administering contracts.

The decision of the court ended in April 2016. The case was one of the longest and most technically complex to come to trial in Scotland in recent years and was held between October 2015 and April 2016. The court sat for 87 days and over 73,000 documents in evidence were lodged in the electronic bundle, including 40 expert reports and 103 witness statements (Doran 2017). The responsibility for the collapse was examined in court and the findings were later appealed by a further three judges. The judge of the first case found that the cause of the failure was "erodible rock" and the other three judges agreed, but, it is argued, that there is evidence for a complex wedge failure at a scale larger than geological mapping. A detailed study of geology and the case may be found in Hencher (2019). As a result, the client lost £130 claim million in this court battle.

This case is given as an example in this book to show how tunnel failures due to uncertainties in geology may cause tunneling disputes in court lasting several years, asking for several thousands of documents in evidence, and tens of expert reports and witnesses' statements.

12.6   CONCLUDING REMARKS

A financier or a bank creditor when making his assessment relies on whether the client has taken a risk-based approach in choosing the tunnel alignment depending on a reliable baseline report. He will also check whether the client has established a risk management framework based on qualiy, and cost or he has used risk registers and got design checkers. The decisions of financiers and insurers are linked to each other. Due to the interrelationships between risk management, code of practice, and insurance issues, these three parameters are handled together in this chapter, giving some examples from different tunneling projects. In this chapter, first, a flowchart indicating the most essential issues/tasks needing to be accurately fulfilled in the frame of the risk assessment and risk management processes during the design and construction stages of tunnel works is discussed in detail. Later risk assessment examples are given from different projects, including a TBM risk classification for tunnels to be excavated close to the major fault zones, and a risk classification for methane migration for TBM excavating under an old coalfield is outlined within this chapter. An example of a risk analysis aimed at the possibility of achieving a project schedule is summarized in order to give an illuminating idea of how to carry out a risk analysis step by step. The chapter terminates with insurance and claim issues.

REFERENCES

Artopoulos, S., 2015. Tunnel-Insurance-Fact-Trends, Advance International Technical Loss adjusters (AITLA). www.imia.com/wp-content/uploads/2015/11/SP26-2015-Tunnel-Insurance-Facts-Trends-November-2015.pdf. Uploaded on April 2022.

Artopoulos., S., 2016. Risk management in TBM tunnelling-what contractors should know when buying insurance cover, in *Proceedings of 2nd International Conference on Tunnel Boring Machines in Difficult Ground conditions, TBM DiGs, Symposium Organized by Turkish Tunnelling Society*, 16–18 November, Istanbul, pp. 228–236.

Aslanbas, A., Bilgin, N., 2020. Excavating metro tunnels with an EPB TBM under risk of methane explosion, in *ITA World Tunnel Congress (WTC 2020)*, Kuala-Lumpur, Malaysia.

Ballantyne, B., 2018. Modern tunnelling risk, article information and share options, https://corporatesolutions.swissre.com/insights/knowledge/Modern-tunnelling-risk.html. Uploaded 4 April 2022.

Bilgin, N, 2016. An appraisal of TBM performances in Turkey in difficult ground conditions and some recommendations. *Tunnelling and Underground Space Technology*, 57, pp. 265–276.

Bilgin, N., Balci, C., Aslanbas, A, 2021. Case studies leading to the management of tunnel fire risks during TBM drives in an old coalfield. *Tunnelling and Underground Space Technology*, 112(2021), p. 103902. https://doi.org/10.1016/j.tust.2021.103902

Bilgin, N., Copur, H., Balci, C., 2013. TBM *Performance Prediction for Uskudar–Umraniye–Cekmekoy–Sancaktepe Metro Tunnels and Risk Analysis*. Report of Investigations, Istanbul Technical University.

Copur, H., Cinar, M., Okten, G., Bilgin, N., 2012. A case study on the methane explosion in the excavation chamber of an EPB-TBM and lessons learnt including some recent accidents. *Tunneling Underground Space Technology*, 27(1), pp. 159–167.

Doran, K., 2017. *Contractor Victorious in Tunnel Collapse Claim, Construction Bulletin*, March 14.

Eskesen, S.D., Tengborg, P., Kampmann, J., Veicherts, T.H, 2004. Guidelines for tunnelling risk management: International tunnelling association, *Working Group No. 2. Tunnelling and Underground Space Technology*, 19, pp. 217–237.

Gaillard, C., Humbert, E., Robert, A., 2013. Application of a methodology for risk management on tunnel project, in *World Tunnel Congress 2013 Geneva Underground – the Way to the Future!*, G. Anagnostou & H. Ehrbar, (eds), Taylor & Francis Group, London.

Guclucan, Z., Meric, S., Algan, M., Palakci, Y., Bilgin, N., Bilgin, A.R., Balci, C., Tumac, D. (2008). The use of a TBM in difficult ground conditions in Beykoz – Kavacik Sewerage Tunnel. Proceedings of the World Tunnelling Congress, Underground Facilities for Better Environment and Safety, Agra, India.

Hencher, S.R., 2019. The glendoe tunnel collapse in Scotland. *Rock Mechanics and Rock Engineering*, 52(2019), pp. 4033–4055. https://doi.org/10.1007/s00603-019-01812-w

ITIG, 2016. *A Code of Practice for Risk Management of Tunnel Works, Prepared by ITG (The International Tunnelling Insurance Group) with the Collaboration of British Tunnelling Society*, p. 28.

Kissel, F.N., 2006. Preventing methane gas explosions during tunnel construction, handbook for methane control in mining. *IC 9486*. Chapter 14, pp. 169–184.

Law Explorer, 2015. Risk in underground construction 05 Oct, Foreign and international Law. https://lawexplores.com/risk-in-underground-construction/

Namlı, M., Bilgin, N., 2017. A model to predict daily advance rates of EPB-TBMs in a complex geology in Istanbul. *Tunneling and Underground Space Technology*, 62, pp. 43–52.

Namlı, M., Cakmak, O., Pakis, I.H., Tuysuz, L., Talu, T., Dumlu, M., Balci, C., Copur, H., Bilgin, N., 2013. *A Methodology of Using Past Experiences in the Performance Prediction of a TBM in a Complex Geology and Risk Analysis*. World Tunnel Congress, Geneva, Switzerland.

Namlı, M., Cakmak, O., Pakis, I.H., Tuysuz, L., Talu, T., Dumlu, M., Şavk, S., Bilgin, N., Copur, H., Balci, C., 2014. *The Performance Prediction of a TBM in a Complex Geology in Istanbul and the Comparisons with Actual Values. Proceedings of the World Tunnel Congress 2014 – Tunnels for a Better Life*. Foz do Iguaçu, Brazil.

Shahriar, K., Sharifzadeh, M., Hamidi, J.K, 2008. Geotechnical risk assessment-based approach for rock TBM selection in difficult ground conditions. *Tunnelling and Underground Space Technology*, 23, pp. 318–325.

Wannick, H.P., 2016. The code of practice for risk management of tunnel works, future tunnelling insurance from the insurers' point of view, *ITA Conference Seoul*, April 25, 2006.

13 Health and Safety in Mechanized Tunneling

13.1 INTRODUCTION

Tunneling is one of the fields with the highest potential for risk and hazards, leading to severe accidents in the entire construction industry. The most common hazards are collapses, gas, water, or fire. Modern tunneling requires competent planning to achieve maximum safety for tunnel workers and, at the same time, to maintain desired production efficiency. All safety aspects must be addressed during the early stage of the planning process. A safety management plan must be concerned with the identification of safety-relevant issues, the definition of parameters to be observed, observation methods, reading frequency, evaluation methods, the definition of warning and alarm levels and criteria, and the definition of contingency measures for each warning level, action plan in case of an alarm, and organization plan of the staff to be involved in the management plan and reporting structure. The target must be "as low a level of risk as practicable possible", so-called ALARP, that is based on risk assessments related to specified acceptance levels. The contractor is given the responsibility for identifying and assessing the risk elements and has to describe and document the complete internal control system in a written report. However, in most tunnels, there is a potential for geological hazards to develop into accidents, or health risks, if they are not recognized during planning and construction, and if appropriate preparations are not made. This applies to the risk of flooding, collapse in unconsolidated zones, crushed or blocky rock masses, and methane emissions. For the different tunneling activities, a detailed job hazard analysis must be carried out where all the potential hazards are identified for each work activity. For each of the job hazards in each construction phase, safety procedures must be identified. Safety control measures are then put in place to minimize or avoid job hazards, i.e., regular safety meetings, briefings, seminars, and audits. The contractor together with the client should conduct a regular review of the safety measures. Corrective measures and improvements must be done, leading to an efficient safety management plan where there are shortcomings.

DOI: 10.1201/9781003358978-13

13.2 THE HEALTH AND SAFETY HAZARDS IN TUNNEL CONSTRUCTIONS

Geological uncertainties create hazards that control the economy and the efficiency of tunneling. Typical geological features that represent hazards to the safety and health in tunneling are defined by Blindheim (2014) as, water under pressure, unconsolidated clayey or sandy zones, rock stress, blocky rock mass, and gas. Water under pressure may cause flooding of the tunnel, the main prevention is a probe drill to localize potential inflow, and pre-grouting and/or drainage will be a definite prevention of the hazard. Unconsolidated zones may cause an immediate cave-in blocking/squeezing the TBM cutterhead, the main signal being water, mud in the probe drill, and the change in probe drilling rate. The prevention may be an injection of fine cement or double-acting polyurethane and umbrella arch as applied in Kargi HEPP tunnel project (Bilgin et al. 2016). Poor confinement or blocky rock mass may create a hazard of damaging the cutters or blocking the cutterhead of a TBM as happened in Göztepe–Üsküdar Tunnel remedial work being the mounting grizzly bars in the cutterhead (Yüksel et al. 2015). The other important geological hazard is the methane emission in the tunnel, warning signals being bubbles in seepage water and a rotten smell of associated gas. Effects of potential consequences may be explosions and delays in work activities as happened in Selimpaşa Tunnel (Copur et al. 2012). Preventive actions for methane explosion will definitely be probe drilling, increased ventilation for dilution, measurements, and monitoring.

13.3 REGULATIONS AND STANDARDS

The Secretary of State for Transport in the UK released "The Road Tunnel Safety Regulations 2007", which is mostly taken into consideration in different countries and publicly available at www.legislation.gov.uk/uksi/2007/1520/made. The readers may find the following items in these regulations: designation of administrative authority, duties of administrative authority, duty to compile accident and fire reports, suspension or restriction of the use of a road tunnel, designation of the tunnel manager, duties of the tunnel manager, designation of the safety officer, duties of safety officers, the appointment of inspection entity, duties of inspection entity, appointment and duties of the technical approval authority, road tunnels already in operation, safety requirements to be met by road tunnels, risk reduction measures, risk analysis, derogation for innovative techniques, application of regulations where the same person undertakes functions of different authorities.

Another important legislation is Directive 2004/54/EC, which the European Parliament and the Council of the European Union have adopted. This directive aims at ensuring a minimum level of safety for road users in tunnels in the trans-European road network. It aimed to apply to all tunnels in the trans-European road network with lengths of over 500 m, whether they are in operation, under construction, or at the design stage. In order to implement a balanced approach and due to the high cost of the measures, minimum safety equipment should be defined, taking into account the type and the expected traffic volume of each tunnel. Directive 2004/54/EC consists mainly of FTA (Fault Tree Analysis) method for event frequency assessment and

ETA (Event Tree Analysis) for event consequences assessment for risk assessment. However, Stefanak et al. (2008) criticized this directive, stating that in the methodology used there is a problem with node selection influencing users' self-rescue. In the newly proposed model they suggest, the most important nodes influencing the user's self-rescue in a tunnel are smoke presence in the first 5 min, automatic ventilation start-up, and user evacuation possibilities, which are affected by emergency exit distance. They insist on solving problems by a model which removes uncertainties using an expert system.

13.4 FIRE

Tunnel fires can have catastrophic consequences in terms of human and structural damage. Beard and Carvel (2011) reported major tunnel fires in 21 countries from 1842 to 2010 with a total fatality of 1,701 (except Salang Tunnel in Afghanistan with fatalities expected to be from 176 to several thousand) and concluded that the analysis done clearly could give valuable lessons for future applications. Due to the importance of this subject, this topic will be treated in detail in Chapter 14.

13.5 FLOOD RESCUE AND ESCAPE

As Sousa and Einstein (2021) noticed, within 64 cases of accidents studied in the worldwide, the percentage of flooding is 12.6%. It is interesting to note that according to Zhu et al. (2022), in the 48 cases of accidents studied in China, the percentage of flooding is 12.4% in the same order as the previous study. However, the consequences of the flooding may be catastrophic in terms of fatalities, delayed time schedules for tunnel construction, and financial losses. As given some examples, in Gerede (Turkey) Tunnel a TBM was lost and the project was delayed almost a year due to groundwater with high water pressure flooding the tunnel and destroying segments and the tunnel, as seen in Figures 13.1 and 13.2 (Bilgin 2016).

An example of a flooding tunnel from heavy rains is Veligonda Tunnel from AMR Project in India. A major problem occurred on 2 October 2009, at the AMR Inlet, when the job site was hit by a 100-year monsoon that flooded the job site and covered the TBM and backup. The equipment was under water for approximately ten days until the water was pumped out. The equipment was completely refurbished by Robbins personnel at the job site. Figure 13.3 shows the recovery of TBM from the flooded Veligonda Tunnel (Harding 2010).

The water inflow accident in the Anshi Tunnel in China is a typical example of tunnel flooding, resulting from unexpected geological conditions (Xinghau 2019). The Anshi Tunnel was excavated by New Australian Tunneling Method, located in the karst area of Yunnan Province in Southwest China. On 26 November 2019, two successive water inflow accidents occurred in the tunnel. An unforeseen water-bearing fracture zone of 15,300 m^3 above the tunnel was ignored due to the lack of detailed prospecting. In the first incident, the inflow point was blocked rapidly by siltation, and only a small amount of mud-rock flowed into the tunnel. However, the current workers carried out the excavation without paying too much attention to the

FIGURE 13.1 Ground water with high pressure destroying the segments in Gerede Tunnel in Turkey. (Bilgin 2016.)

FIGURE 13.2 Ground water in Gerede Tunnel with high pressure destroying the tunnel. (Bilgin 2016.)

FIGURE 13.3   Recovery of TBM from flooded Veligonda Tunnel. (Harding 2010.)

accident. Finally, an extremely huge amount of mud-rock poured into the tunnel, and the avalanche flow of mud-rock destroyed everything in the tunnel, resulting in 12 casualties, and the loss of all construction machinery.

One of the most significant flooding accidents happened in the tunnels of the Storebælt fixed link project involving a twin bored railway tunnel 8 km long which connects the islands of Sprogo and Seeland in Denmark. The causes and the impact of this important accident on project scheduling, taking into consideration the geotechnical aspects of the project, are explained by Anagnastou (2014), Muir Wood Lecturer at World Tunnelling Conference, Brazil, 1994. The tunnel was constructed from 1990 to 1997 using four EPB shields of 8.75 m in diameter. There were delays and cost overruns in the tunnel construction. The plan was to open it in 1993, giving the trains a head start of three years over road traffic, but train traffic started in 1997 and road traffic in 1998. During construction in October 1991, the sea bed gave way and one of the tunnels was flooded. The quantity of water flowing in increased rapidly to 4 m^3/sec and led to the flooding of the tunnel. The water continued to rise and reached the end at Sprogø, where it continued into the (still dry) other tunnels. The water damaged two of the four TBMs, but no workers were injured. It was possible to dry out the tunnels only by placing a clay blanket on the sea bed. The remedial work took about eight months to complete.

One of the most important tunnel flooding worth mentioning is Pavoncelli Bis Tunnel which started in 1990 and stopped after about 580 m (5% of the route) due to uncontrollable groundwater flows that imposed work abandonment. The tunnel was completed in October 2017.

13.6 HYPERBARIC INTERVENTION

In pressurized tunneling as in EPB tunnel drives, intervention in compressed air conditions is sometimes necessary to change the cutters and to check the cutterhead in unstable tunnel faces. Working in compressive air increases the risks of working in tunnels in respect of occupational health and safety. The workers working in a compressed air environment should pass a special training course and pass a medical check-up realized by a doctor competent in hyperbaric medicine and have hyperbaric medical fitness. All TBMs intended to work in difficult conditions should have a man lock allowing workers to move and sit down having specific dimensions. The lock must be equipped with first aid equipment, an internal water spray system, a hyperbaric fire extinguisher, and a communication system. Medical locks should be provided when working pressure exceeds 0.7 bar and are normally used for therapeutic treatment. Under normal conditions, a maximum working pressure of 3.6 bar is allowed in the working chamber. Before starting in compressed air, an emergency plan should be prepared and posted in suitable places. In the working chamber and on the tunnel face, the air quality should be constantly monitored and additional air must be supplied if necessary. Compression and decompression operations can adversely affect the health of the workers, so they must be carried out with the utmost care. Decompression should be done gradually in accordance with prescribed rules. Normal working time is limited to 4 h. Decompression illness can occur anytime during the first few hours after leaving the lock and the diver should not participate during the first hours of work requiring extensive physical efforts (ITA report no 001, 2008). ITA published another detailed report no 010 (2015), which gives detailed information on legislation, standards, plant, equipment, gas supply, and related topics on hyperbaric working conditions. The readers are advised to read these two reports to have more ideas about this important subject of tunneling concurring in difficult ground conditions.

One of the typical examples working in hyperbaric conditions is Istanbul New Airport Metro tunnels bored through a frequently changing and complex geology and had to be maintained and repaired under hyperbaric conditions as it was lying 75 m under a heavily used highway heading to the new airport. In this project, compressed air diving operations were realized for intervening in the cutterhead. The selected diving team was not only experienced in compressed air TBM tunnel operations but also in saturation diving differing between 6.0 and 11.8 bars in Turkey and China (Erboylu et al. 2020).

Following machinery and tools were made available for the operation. To maintain the air pressure inside the chamber along the diving operation in addition to the current air compressor of 13.42 m³/min capacity, another same capacity oil-injected screw compressor was provided to the TBM. For supplying air to the pneumatic hand tools used inside the excavation chamber, an electric air compressor with a capacity of 600 L/min was also used. Also for emergency situations like power cuts, a diesel generator was provided. Welding machines, grinding machines as well as pneumatic torque tools, and chain hoists were available. Water jet with up to 500 bar pressure for cleaning the cutting tools housings and back side of the cutterhead was provided. Projectors and torches working with low voltage, namely 24 volts to prevent possible electric shocks, were the tools to lighten the excavation chamber. The pressure inside the excavation chamber was detected via the sensor on the airline connected to Samson unit system. Manlock's inner pressure was increased up to the aimed pressure level and controlled any pressure loss or air leakage (Erboylu et al. 2020).

Each time two of the divers entered the manlock, under the diving operator's supervision, they got pressurized to the excavation chamber's current pressure when the aimed pressure value was digitally entered into the Samson's pressure control unit. Pressurized air is derived from the air compressor located at the backup gantry.

As seen in Table 13.1, between 20 and 28 March 2019, a total of 45 divings were realized. Divers changed 25 discs with 17 of them missing, 4 of them broken, and 3 of them worn over 25 mm. After changing disc cutters, the performance of TBM increased considerably. A diver working in the hyperbaric condition in the tunnel face and in the manlock is seen in Figures 13.4 and 13.5

13.7 DUST

Geologic formations containing silica and asbestos may cause severe health and safety environmental problems. Dust is produced especially when a tunnel is being constructed using an open-type TBM in hard rock, drilling the rock, and transporting the muck, which is an important factor that affects the operating environment. In such cases, it is definitely necessary to design an effective ventilation dust removal system to suppress the diffusion of dust particles as studied by Zhou et al. (2020).

Bakke et al. (2001) reported a research study on personal exposures to dust and gases among 189 underground construction workers grouped as drill and blast crew, shaft-drilling crew, TBM crew, shotcrete operators, support workers, concrete

TABLE 13.1

TBM Hyperbaric Intervention Beneath a Highway in New Istanbul Airport Metro Project

Total Dive Number	Total Diving Duration	Average Working Duration per Dive	Average Decompression Time per Dive	Average Working Pressure
45	7,426 min	86 min	80 min	3.30 bar

Source: Erboylu et al. (2020).

FIGURE 13.4 Diver working in hyperbaric condition in a tunnel face. (Erboylu et al. 2020.)

FIGURE 13.5 Diver working in hyperbaric condition in man lock. (Erboylu et al. 2020.)

workers, and electricians. Outdoor tunnel workers were included as a low-exposed reference group. The highest geometric mean (GM) exposures to total dust (6–7 mg/m^3) and respirable dust (2–3 mg/m^3) were found by the shotcrete operators, shaft drillers, and TBM workers. Shaft drillers and TBM workers also had the highest GM exposures to respirable alpha-quartz (0.3–0.4 mg/m^3), which exceeded the Norwegian occupational exposure limit of 0.1 mg/m^3. Similar work repeated by Bakke et al. (2002) revealed that job groups with the highest GM total dust exposure were shotcrete operators (6.8 mg/m^3), TBM workers (6.2 mg/m^3), and shaft drilling workers (6.1 mg/m^3). The lowest exposed groups to total dust were outdoor concrete workers (1.0 mg/m^3), electricians (1.4 mg/m^3), and support workers (1.9 mg/m^3).

One compiled study on construction dust, causes, effects, and remedies was carried out by Subramanian and Abhyankar (2019) with the following summary. Silica dust is created when workers are dealing with rocks containing crystalline silica, such as sandstone, granite, and rhyolite. Silica dust is more injurious to health and hence several countries have regulations to limit it. Inhaling respirable crystalline silica can cause diseases like silicosis and lung cancer. In most cases, these diseases occur after several years of exposure and hence are predicted only during the critical stages. Every year in Great Britain, over 500 construction workers are believed to die from lung cancer caused by silica dust alone. The Occupational Safety and Health Administration of the USA developed a standard that limits worker exposure to silica to an average of 50 µg/m^3 over 8 h. In the UK, the Health and Safety Executive has set a maximum exposure limit for silica dust of 0.3 mg/m^3 (averaged over 8 h).

13.8 GASES: LUCK OF OXYGEN

Various hazardous gazes may exist within the tunnel. Adequate ventilation is requested to remove the polluted air and also to ensure temperatures of not more than 400 °C dry and 290 °C in the working place. It is demanded that the concentration of various gases in the atmosphere inside the tunnel should be as follows as defined in NSCI (2016):

a. Methane should be measured close to the roof and not exceed 0.5% at any place inside the tunnel.
b. Carbon monoxide should be less than 0.005%.
c. Carbon dioxide should be less than 0.5%.
d. Nitrogen fumes should be less than 0.0005%.
e. Hydrogen sulfide should be less than 0.001%.
f. Aldehyde should be less than 0.0002%.

It is advisable that the oxygen content in the tunnel atmosphere should not be less than 19%. One of the most important accidents occurring in tunnels is due to methane explosions. Studies by Copur et al. (2012) and Bilgin et Al. (2016) summarize accidents in this respect, project name, the result of the accidents, and the causes of the accidents are given in Table 13.2.

TABLE 13.2
Methane Explosions Occurred in Tunnels in Recent Years

Project	Tunnel Length	TBM Diam. Date	Accident Type	Result of Accident	Causes of Accident	Ref
Los Angeles Water Tunnel,	8.85 km	5.5 m 1971	Methane explosion	17 died	Gas detector didn't function	Proctor (2002)
Higashimurayama, Japan	–	6 m 1978	Methane explosion	9 died, 2 injured	Power supply did not shut off	Kitajima (2010)
Abbeystead Valve House, UK	–	–	Methane explosion	16 died, 10 injured	Methane in groundwater	Lockyer and Howcroft (1997)
Aqueduct Tunnel in Carsington, UK	8.5 km	2.4 m 1987	Methane in tunnel	No fatality or injury	Methane in groundwater	Pearson et al. (1989)
Tunnel, Huyiki Tokyo	–	6.5 m 1993	Methane explosion	4 died, 1 injured.	Staff didn't notice the alarm	Kitajima (2010)
Mill Creek Tunnel, USA	4.65 km	7.8 m 2004	Methane in tunnel air	No fatality or injury	Natural gas field	Schafer et al. (2007)
Zagros Tunnel, Iran	26 km	6.73 m 2009	H_2 and methane	No fatality or injury	Oil and gas bearing-basins	Shahriar et al. (2009)
Selimpasa Tunnel, Istanbul	10.63	3 m 2010	Methane explosion	8 injured	Methane in ground water	Copur et al. (2012)
Silvan Irrigation Tunnel, Turkey	10.31	7.85 m 2015	Methane combustion	13 injured	Natural gas reservoir	Bilgin et al. (2016)

Sources: Copur et al. (2012) and Bilgin et al. (2016).

At higher altitudes, workers' health and efficiency are significantly influenced by low oxygen concentration and low temperature. It is reported by Guo et al. (2016) that in China the number of tunnels opened at higher altitudes than 3,000 m is increasing steadily year by year. Realizing the impacts of a high-altitude environment on human bodies, they analyzed the relationship between labor intensity and oxygen consumption in high-altitude areas and determined the critical oxygen-supply altitude values for tunnel construction. In high-altitude tunnels between 2,000 and 4,000 m, the saturation of blood oxygen drops by about 12%, and the heart rate and arterial pressure increase by about 12% compared with the data at sea-level locations. In these cases, the workers need to rest more frequently and working efficiency is significantly lower. In ultra-high-altitude tunnels, the saturation of blood oxygen drops by about 20%, and the heart rate and arterial pressure increase by about 18% and 17% (Wu et al. 2019).

Most railway tunnels in the Qinghai–Tibet lie in areas over altitudes of 4,000 m and the lower oxygen concentration in the air becomes the key problem for workers' health and safety, as Liu et al. reported (2010). Their study is based on concentrating oxygen by developing a new technology called "a pressure swing adsorption method". In the high-altitude areas in the Fenghuoshan Tunnel, the comparison of the physiological effects on the construction workers before and after oxygen supply using this new technology was very positive.

13.9 ACCIDENTS, GROUND INSTABILITY, TRIP AND FALL

The most comprehensive work on accidents during tunnel construction was done recently by Sousa and Einstein (2021). They worked on 206 cases that occurred during construction. The study of the accidents in the database made it possible to identify and categorize accidents into different types, typical causes, and consequences, as well as the identification of the scenarios in which these events (accidents) are more likely to occur. The analysis of the database shows that unexpected ground conditions are often the main reason for tunneling accidents during construction. Despite recent efforts made to improve existing tunneling technologies, forecasting ground conditions in tunneling remains the most challenging task because of significant uncertainties related to geology. Accident-type distribution and the distribution of accidents according to the construction method are given in Tables 13.3 and 13.4.

Sousa and Einstein (2021) concluded that in 64 cases for which data on delays were available, most of the delays due to the accidents ranged between 0 and 7 months, with an average of around 6 months. Since the 1990s, there have been several great losses involving tunnels in urban areas causing, in some cases, repairs costing up to US$100 million. From the 1990s to the early 2000s, CAR (contractors all risks) insurers have suffered losses totaling up to more than US$750 million in property damage only (Landrin et al. 2006). Sousa and Einstein (2021) also concluded that common to many accidents described was the fact that the main reported causes were unpredictable geological conditions (external cause), whether they consisted of fault zones (and their extent), other weak zones, or groundwater presence.

Another interesting study on major tunnel construction accidents that happened in China between 2010 and 2020 was carried out by Zhu et al. (2022). Their study is

TABLE 13.3
Accident-type Distribution in the Database

Event	Number of Events	Percentage
Collapse	78	41
Daylight collapse	52	28
Excessive deformation	19	10
Rockfall	12	7
Flooding	12	6
Specific location	10	5
Rock burst spalling	2	1
Other	4	2

Source: Sousa and Einstein (2021).

TABLE 13.4
Distribution of Accident Type According to the Construction Method

Event	Conventional Method	TBM
Collapse	33	45
Daylight collapse	34	11
Excessive deformation	12	11
Rockfall	8	7
Flooding/high water ingress	5	18
Specific location	4	3
Rock burst spalling	2	2
Other	2	3

Source: Sousa and Einstein (2021).

important in a way that it shows that driving tunnels with TBM is safer than the conventional tunneling method, with the ratio of fatalities to incidents being 6.1 in drill and blast tunneling method and 5.6 in mechanical excavation with TBM as seen in Figure 13.6.

Accident types, incident numbers, and fatalities in tunnel construction in China between 2010 and 2020 are seen in Table 13.5 (Zhu et al. 2022). As seen from this table, the highest fatality/incident ratio is for an explosion as an accident type.

13.10   ELECTROCUTION

The harsh conditions on many tunneling sites can damage electrical equipment and cables and reduce their lifespan. Most tunnels are wet or damp, providing a

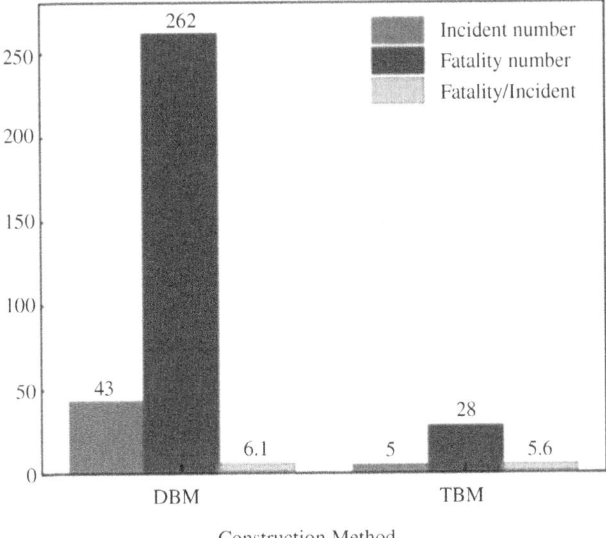

FIGURE 13.6 Major tunnel accidents and deaths numbers in TBM and drill and blast method. (Zhu et al. 2022.)

TABLE 13.5
Accident Type, Incident Number, and Fatalities in Tunnel Construction in China Between 2010 and 2020

Accident type	Incident Number	Fatalities	Fatalities/Incident
Collapse	31	172	5.55
Explosion	8	68	8.5
Water inrush	6	38	6.33
Suffocation/fire	2	9	4.50
Rock burst	1	3	3.00

Source: Zhu et al. (2022).

perfect ground for short circuits resulting in frequent electrocutions. As a result, serious injuries may occur from exposure to electrical hazards on tunneling sites. It is recommended that all necessary materials be properly grounded and that the switches are located on high ground and these be properly grounded. Electric shock is the main risk and electrocution is the most fatal electrical hazard. However, it should be noted that the risk of electrocution in TBM tunneling is less than in the conventional excavation method (Couto et al. 2018).

13.11 CONCLUDING REMARKS

Tunneling is one of the fields with the highest potential for risk and hazards, leading to severe accidents in the entire construction industry. This aspect of tunneling necessitates setting up a health and safety plan which should form the basis for the identification and management of all health and safety risks arising from the works. Geological uncertainties create hazards such as water under pressure, unconsolidated clayey or sandy zones, rock stress, blocky rock mass, and gas which control the economy and the efficiency of tunneling. It is reported that since the 1990s, there have been several great losses involving tunnel collapses in urban areas causing, in some cases, repairs costing up to US$100 million. From the 1990s to the early 2000s, insurers have suffered losses totaling up to more than US$750 million in property damage only. Various hazardous gases may exist within the tunnel which affect the health of the workers. They should be monitored carefully and an efficient ventilation system is inevitable as remedial work. Methane leading to an explosion in the tunnel or even in the pressure chamber of a TBM is important and needs to be handled carefully. It is reported that from the accidents that happened during tunnel construction, the percentage of flooding is around 12%. However, the consequences of the flooding may be catastrophic in terms of fatalities, delayed time schedules for tunnel construction, and financial losses. In pressurized tunneling as in EPB tunnel drives, intervention in compressed air conditions is sometimes necessary to change the cutters and to check the cutterhead in unstable tunnel faces. Working in compressive air increases the risks of working in tunnels in respect of occupational health and safety. All these hazards and accidents may be minimized with a well-executed geological baseline report and with a well-designed health and safety plan.

REFERENCES

Anagnastou, G., 2014. Some critical aspects of subaqueous tunnelling, Muir Wood Lecturer. *ITA-AITES World Tunnelling Congress*, Brazil, pp. 1–20.

Bakke, B., Stewart, P., Eduard, W., 2002. Determinants of dust exposure in tunnel construction work. *Comparative Study Applied Occupational and Environmental Hygiene*, 17(11), pp. 783–96. doi:10.1080/10473220290096032

Bakke, B., Stewart, P., Ulvestad, B., Eduard, W., 2001. Dust and gas exposure in tunnel construction work. *American Industrial Hygiene Association*, 62(4), pp. 457–465. doi:10.1080/15298660108984647

Beard, A., Carvel, R., 2011. *Handbook of Tunnel Fire Safety*, 2nd edition. ICE Publishing, London.

Bilgin, N., 2016. An appraisal of TBM performances in Turkey in difficult ground conditions and some recommendations. *Tunnelling and Underground Space Technology*, 57, pp. 265–276.

Bilgin, N., Copur, H. and Balci, C., 2016. *TBM Excavation in Difficult Ground Conditions. Case Studies from Turkey*. Ernst & Sohn, Berlin.

Blindheim, O.T., 2014. *Geological hazards, causes, effects, and prevention, health and safety in Norwegian tunnelling*, Norwegian Tunnelling Society Publication No.13, pp. 23–29.

Copur, H., Cinar, M., Okten, G., Bilgin, N., 2012. A case study on the methane explosion in the excavation chamber of an EPB-TBM and lessons learned including some recent accidents. *Tunneling and Underground Space Technology*, 27(1), pp. 159–167.

Couto, J.P., Camões, A., Tende, M.L., 2018. Risk evaluation in tunneling excavation methods. *REM – International Engineering Journal*, 71(3), pp. 361–369. https://doi.org/10.1590/0370-44672017710115

Directive 2004/54/EC, 2004 of the European Parliament and of the Council on Minimum Safety Requirements For Tunnels, *in The Trans-European Road Network*, Brussels, 29 April 2004.

Erboylu, U., Acun, S., Senyurt, T., Bayram, D., 2020. TBM hyperbaric intervention Beneath a highway in New İstanbul Airport Metro Project, *ITA-AITES World Tunnel Congress*, Kualalumpur, Malaysia, 15–21 May.

Guo, C., Xu, J., Wang, M., Yan, T., Yang, L., Sun, Z., 2016. Study on oxygen supply standard for physical health of construction personnel of high-altitude tunnels. *International Journal of Environmental Research and Public Health*, 13, p. 64. doi:10.3390/ijerph13010064

Harding, D., 2010. Tunnel boring machines used for irrigation in Andhra Pradesh, India, *ITA-AITES World Tunnel Congress*, 14–20 May, Vancouver, Canada.

ITA report no 001, 2008. *Guidelines for Good Occupational Health and Safety Practice in Tunnel Constructions*. November, p. 44.

ITA report no 010, 2015. *Guidelines for Good Working Practice in High Pressure Compressed Air*. April, p. 36.

Kitajima, M., 2010. Methane gas explosion hazard during construction of tunnel for agriculture. www.shippai.org/fkd/en/cfen/CD1000099.html (accessed on 05.03.10).

Landrin, H., Blückert, C., Perrin, J.P., Steve Stacey, S., Stolfa, A., 2006. ALOP/DSU coverage for tunneling risks?. *The International Association of Engineering Insurers*. 39th Annual Conference.

Liu, Y.S, Wu, T.Y., Ding, S.Q., Liu, W.H., Hou, Q.W., Feng, J.X., Le, K., Zhang, H., Li, Y.L, 2010. Oxygen concentrating and application for railway tunnel construction in high altitude area and it's physiological effects on the construction workers, *4th International Conference on Bioinformatics and Biomedical Engineering, Chengdu, China*. doi:10.1109/ICBBE.2010.5517173

Lockyer, J.W., Howcroft, A., 1997. The Abbeystead explosion disaster. *Annals of Burns and Fire Disasters*, 10, pp. 1–4, September.

NSCI, 2016. Safety in tunnelling excavation, National Safety Council, India, p. 12. https://vidyutbodha.files.wordpress.com/2016/04/6-1safety-in-tunneling-excavation.pdf, uploaded in April 2022.

Pearson, C.F.C., Edwards, J.S., Durucan, S., 1989. Methane occurrences in the Carsington Aqueduct tunnel project – A case study. *Proceedings of the Rapid Excavation and Tunneling Conference*, pp. 176–195.

Proctor, R.J., 2002. The San Fernando tunnel explosion. *Engineering Geology*, 67, pp. 1–3.

Schafer, M., Pintabona, R., Lukajik, B., Kritzer, M., Janoska, S., Switalski, R., 2007. Gas mitigation in the Mill Creek Tunnel. *Proceedings of the Rapid Excavation and Tunneling Conference*, pp. 168–175.

Shahriar, K., Rostami, J., Hamidi, J.K., 2009. TBM tunneling and analysis of high gas emission accident in Zagros long tunnel, in *ITA World Tunnel Congress (WTC 2009)*, Budapest-Hungary, pp. 171–172.

Sousa, R.L., Einstein H., 2021. Lessons from accidents during tunnel construction. *Tunnelling and Underground Space Technology*, 113 (2021) 103916, p. 28.

Stefanak, J., Juraj Spalek, J., Kallay, F., 2008. Road tunnels safety according to European Legislation. *Transport Problems (Problemy Transportu)*, 4(1), pp. 65–70.

Subramanian, N., Abhyankar, E.V., 2019. Construction dust causes, effects, and remedies, *NBM&CW,* pp. 148–160.

The Road Tunnel Safety Regulations 2007, No 1520, uploaded on 14 August 20022. www.legislation.gov.uk/uksi/2007/1520/

Wu, P., Yang, F., Zheng, J., Wei, Y., 2019. Evaluating the highway tunnel construction in Sichuan Plateau considering vocational health and environment. *International Journal of Environmental Research and Public Health,* 16(23), p. 4671. doi:10.3390/ijerph16234671

Xinghau, W., 2019 Tunnel rescue, disasters, *China Daily,* 28 November.

Yuksel, A., Arioglu, E., Bilgin, N., 2015. Mechanism of roof collapses in front of a TBM in a complex geology in Kadikoy-Metro Kozyatagi Metro Tunnels. *International Conference on Tunnel Boring Machines in Difficult Grounds (TBM DiGs),* 18–20 November, Singapore.

Zhou, W., Nie, W., Liu, X., Zhou, C., Wei, C., Liu, C., Liu, Q., Yin, S., 2020. Optimization of dust removal performance of ventilation system in tunnel constructed using shield tunneling machine. *Building and Environment,* 173, 106745.

Zhu, Y., Zhou, J., Zhang, B., Wang, H., Huang, M, 2022. Statistical analysis of major tunnel construction accidents in China from 2010 to 2020. *Tunnelling and Underground Space Technology,* 124 (2022) 104460, p. 14. https://doi.org/10.1016/j.tust.2022.104460

14 Management of Tunnel Fire Risks During TBM Drives

14.1 INTRODUCTION

Tunnel fires can have catastrophic consequences in terms of human and structural damage. Beard and Carvel (2011) reported major tunnel fires in 21 countries from 1842 to 2010 with a total fatality of 1,701 (except Salang Tunnel in Afghanistan with fatalities expected to be from 176 to several thousand) and concluded that the analysis done clearly could give valuable lessons for future applications. Tunnel safety has increased significantly after previous tunnel fires, such as the Mont Blanc Tunnel fire (France, 1999, 39 fatalities), Tauern Tunnel fire (Austria, 1999, 12 casualties), Gotthard Tunnel fire (Switzerland, 2001, 11 casualties), and Yanhou Tunnel fire (China, 2014, 40 fatalities). This is due to research works realized and cares taken to mitigate future fire risks (Cafaro and Bertola 2009). Much of the research studies carried out in the past on tunnel fires are related to completed tunnels, and there are only a few works on tunnel fires during construction or repair (Lönnermark et al. 2010). Therefore, this chapter is aimed at tunnel fires that occurred during the construction process.

14.2 THE CAUSES OF TUNNEL FIRES

14.2.1 Hydraulic Oil

The major mining disaster, which occurred on 8 August 1956 at the Bois du Cazier/ Marcinell a coal mine in Belgium, was a key point in using fire-resistant oils in the underground. Smoke and carbon monoxide spread down the mine, killing 274 miners trapped by the fire. This accident revealed the fact that underground fluid storage poses an increased fire risk and may serve as a fuel source during a fire (Totten and De Negri 2011). Bickel et al. (1996) emphasized that only approved fire-resistant hydraulic fluids should be used in hydraulically operated equipment. Oil-filled transformers should not be used underground unless they are in a fire-resistant enclosure. Although it is an apparent necessity to use fire-resistant hydraulic oils underground, unfortunately on 11 November 2000, in Austria, in Kitzsten-horn funicular tunnel, a fire happened in the driver's cab, due to hydraulic oil leakage into

DOI: 10.1201/9781003358978-14

the heater. The fire continued for 3 h, and the train was completely burnt and 155 people died (Lönnermark 2005; Beard and Carve 2011).

On 11 June 1994, a fire occurred in the TBM under Store Baelt in Denmark, in the railway tunnel approximately 2 km from the Zealand side (Ingason et al. 2010). The fire fuel was mainly hydraulic oil from the TBM power system which was distributed at high pressure as an oil spray. The workers did manage to escape to the parallel tunnel through a crossing without any casualties (Ingason et al. 2010). They also emphasized that a fire-resistant hydraulic fluid should be mandatory in stationary construction vehicles working underground. Ingason (2008), in his article on the state of the art of tunnel fire research, gives an overview of tunnel fire research. The European standard for fire-resistant fluids is EN16191, which is mainly based on the British standard mostly derived from the mining industry protocols that have a long tradition in the UK. Although the necessity of using fire-resistant fluids in tunnels is obvious from past experiences, it is not compulsory. This rule was strictly followed in the Avrasya/Eurasia Tunnel in Istanbul, which is a double-deck tunnel of 13.7 m of excavation diameter connecting the Asian and European sides. However, we do not see this serious rule applied in the other mechanized tunnels in Turkey and all over the world.

14.2.2 BELT CONVEYORS

Another source of underground fire leading to very severe casualties is a belt conveyor fire due to poisonous gas. Conveyor belts used in the USA and many other countries' underground coal mines are required to be flame resistant. However, the conveyor belt is also not required to be flame resistant for use in underground metal and non-metal mines (Mitchel et al. 1967; Apte 2006; Francart 2006; Hansen 2009; Smith and Thimons 2010). On 13 May 2014 around 2:45 p.m. fire had broken in Soma coal mine. This was Turkey's worst-ever mining accident: 301 miners including mining engineers died, some burnt alive, others suffocating. According to the witness statements in the mine, the smell of burned belt and cable was felt by the workers. From the witnesses, it was obvious that the reason for the disastrous accident was a belt fire. Soma accident revealed the fact that belt fires may cause catastrophic fire accidents ending with several fatalities even reaching several hundred (Bilgin et al. 2021).

Another example of a belt fire is from a copper mine where 19 workers died and 7 were injured in a belt fire in Kure/Turkey copper mine on 8 September 2004. Welding operations done next to the vertical belt flexowell initiated the fire. Later, the fire spread to the horizontal belt. The copper ore was being transported to the surface by a vertical belt of 150 m in length. The cause of death and the injuries of the workers were due to asphyxiation from harmful gases generated from the belt fire (Bilgin et al. 2021).

14.2.3 METHANE AND EXPLOSION INSIDE THE EXCAVATION CHAMBER OF AN EPB-TBM

Methane (CH_4) is known as a colorless, odorless, and non-noxious gas. Methane, contrary to what is known, is not an explosive gas, but it is a flammable gas. Since the density of methane (0.716 kg/m^3) is lower than the density of air (1.293 kg/m^3), it accumulates at the crown of a tunnel. When it mixes with enough air, it becomes flammable and if the burning is in a closed environment an explosion effect is seen with high pressure and temperature. When the concentration of methane within the air is between 5% (lower limit) and 15% (upper limit), it becomes flammable and, in a closed area, becomes explosive due to sudden combustion. A source of temperature about 650–750 °C is required to create an explosion. It cannot be flammable but can cause asphyxiation due to a lack of oxygen at over 15% concentrations. It just flares or burns at under 5% concentration. Methane can be easily emitted through geological discontinuities, joints, and pores of the ground, as reported by Doyle (2001).

A methane explosion inside the excavation chamber of an EPB-TBM is a very rare case and was not reported before in the literature as the authors' best knowledge. A methane explosion occurred inside the excavation chamber of an EPB-TBM in Silivri–Istanbul on 20 May 2010 (Copur et al. 2012). It is considered that the explosion due to accumulated methane in the pressure chamber of TBM was initiated by sparks created by friction between the screw conveyor and its casing. The explosion forced the muck in the chamber to be blown through the screw conveyor and out of the discharge door (Figure 14.1).

The explosion caused the extrusion of 3–4 m^3 of muck through the screw conveyor into the tunnel in which ten personnel were working and all of them were injured due to the fire that followed the muck blow-out. The TBM operator, who was closest to the discharging door of the screw conveyor, was severely burnt. Four labors working around the segment erector were moderately burnt, while the other five labors working outside the rear shield and close to the shaft KT-4 were slightly injured after being knocked over by the pressure of the explosion. The eight personnel were hospitalized. Electricity was shut down as a result of the accident, which made the rescue efforts quite difficult. This accident clearly showed the importance of site investigations and baseline reports. Any possible gas emission along the tunnel alignment should be investigated by suitable techniques and reported in baseline reports. Methane migrating from fracture zones (Figure 14.2) accumulated in the excavation chamber of EPB-TBM and the explosion occurred due to sparking caused most probably by contact between the screw conveyor and its casing.

The EPB excavation chamber is usually full of muck and foam. It is considered that the amount of air necessary for methane explosion came from the foam. It is strongly advised that future research studies should be focused on developing foam functioning without air in gassy grounds. TBM used in this tunnel did not have any automatic gas measuring device inside the working chamber and in critical parts of the TBM. This case study shows also that bentonite injection in the working chamber of EPB decreases methane emission entering fracture zones.

14.2.4 Methane Problem in Pavoncelli Tunnel in Italy

The Pavoncelli Bis Project was an extremely challenging, complex, and difficult tunnel. During the initial phases of excavation, the presence of methane proved to be quite constant and larger than foreseen. This has required quick adaptations to the

FIGURE 14.1   Extruded muck under discharging door of the screw conveyor. (Copur et al. 2012.)

boring machine together with a newly studied safety procedure, allowing excavation in the presence of gas as described by D'Angelis et al. (2019). The following innovative solutions are described by Bandini et al.(2019). The "TBM – back-up – finished tunnel" system is divided into five homogeneous volumes. Compartmentalization of screw conveyor, muck discharge and air–methane mixture aspiration into dedicated pipes and ventilation, and air movement in the shield and along the conveyor belt were realized. Methane drainage from the excavation chamber and the screw conveyor, improved sealing systems against methane flow from TBM shield precast segments, improved methane monitoring system and safety procedures with specific

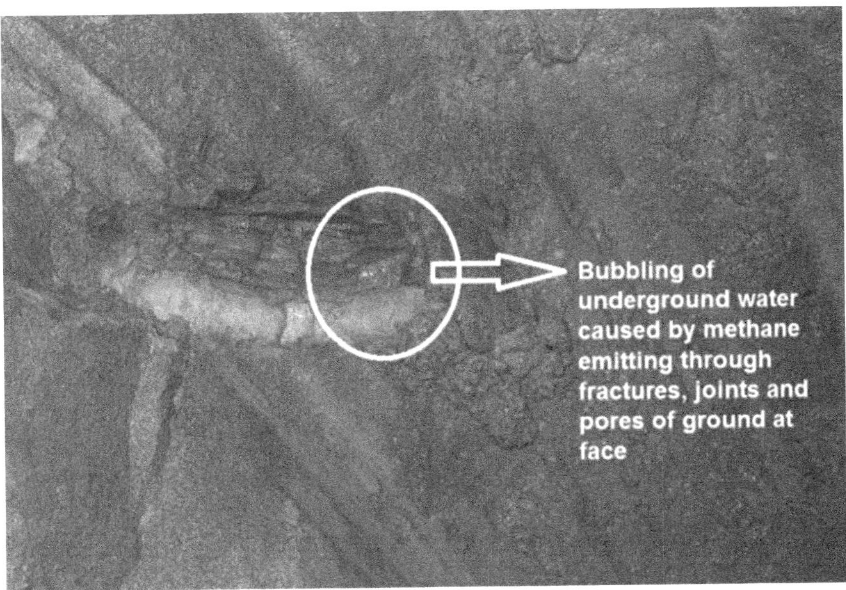

FIGURE 14.2 Methane emission through the fractures and joints in fault zone at the face. (Copur et al. 2012.)

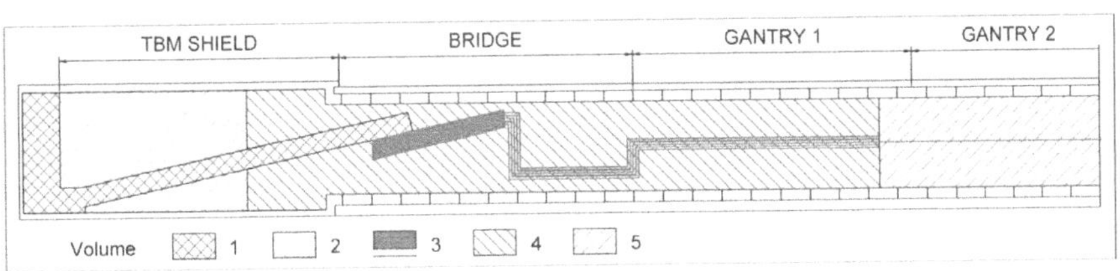

FIGURE 14.3 TBM – back-up – lined tunnel compartmentalization into homogeneous volumes. (Bandini et al. 2019.)

operating procedures for firedamp risk prevention were the main precaution taken during TBM tunneling.

Five homogeneous volumes according to the presence of ignition sources and potential explosive mixtures are given in Figure 14.3. For each volume, early warning and warning threshold levels were defined (Table 14.1). Due to the rigid application of specific safety procedures, the tunnel construction was completed with zero firedamp-related injuries.

14.2.5 Natural Gas Deposits

The fire generated in TBM in Silvan Tunnel in Turkey is a typical example of this type of accident. The basic aim of the Silvan irrigation project is to construct tunnels

TABLE 14.1
Early Warning and Warning Thresholds in the Different Volumes

Volumes	Early Warning	Warning
V1M	14% LEL	20% LEL
3	80% LEL	–
V4	14% LEL	20% LEL
V5	3% LEL	7% LEL

Source: Bandini et al. (2019).

Note: LEL is lower explosion limit, and V1M indicates Volume 1 during maintenance operations and access to excavation chamber.

between the Silvan Dam and agricultural lands in Diyarbakir, in the South-East of Turkey. A double-shield TBM of 7.8 m diameter was used for excavation. Excavation of the 4,668 m tunnel was completed on 21 April 2015 when the gas flaming accident occurred at the chainage 18 + 447 km (Inal and Inal 2015). After the gas flaming incident in the tunnel, it was discovered that the Turkish Petroleum Corporation (TPO) had already made some site investigations along the tunnel and found a natural gas reservoir with a capacity of 24,000 m³/day around 300 m away from the accident area. But neither the project owner nor the contractor had known about this risky situation. The findings of TPO indicate that the natural gas reservoir is of anticline type and had some tensile fractures due to the folding of the surrounding rock formations. These tensile fractures reached up to the tunnel area (Bilgin et al. 2016). Totally 36 personnel were working in the tunnel during the gas flaming on 21 April 2015 at around 9:40 a.m. Thirteen of the personnel were burnt due to the gas fire, of whom five were seriously burnt. The fire/flame/flare engulfed the tunnel starting from the face area through the crown of the tunnel. All of the equipment used in the tunnel were not ex-proof and the gas flaming extensively damaged the TBM.

14.2.6   Ground Contaminated by Gasoline

During site investigations of Bakirkoy–Kirazli Metro Line, Istanbul, it was discovered that an area close to the portal of the tubes was contaminated by leakage of petroleum product, gasoline, coming from a petrol station which is situated nearby to tunnels (Yazici et al. 2019). As the gas station authorities declared, fuel oil contamination was also discovered under another gas station approximately 60 m northeast of the present gas station. Within a 100 m radius of the working site, there were residential apartments, parks, four schools, and a big shopping mall. During the geotechnical studies, it was observed that the soil around the metro line was permeable; thus, it was concluded that the leakage around the gas station had contaminated and spread through the surrounding area. According to the National Institute of Occupational

Safety and Health, the lower exposure limit to gasoline for a human being is 300 ppm for 8 h, short-term exposure is 500 ppm, and it is easily liable to explosions and fire. For that reason, to keep the hydrocarbon threshold below the exposure limits and explosion limits, several precautions and different management tactics for tunnel fire risk were implemented.

First, bentonite was used in order to dilute the concentration of the hydrocarbons within the muck. This application was abandoned after a few attempts since bentonite mixed with other clay minerals found in the excavated material caused clogging of the disc cutter and affected the performance of TBM. Later, gas sensors, of $1 \times O_2$, $1 \times CO$, $1 \times CO2$, and $2 \times CH_4$, were installed and integrated into the control system of TBM. In case these sensors detect gases above the programmed limit, TBM stops automatically. Gas measurements were taken instantaneously from the excavation chamber through a hose line connected to the bulkhead where methane and hydrocarbon gas sensors were fixed. Gas in the chamber runs through a flow meter and reaches the gas sensor. A flow meter is adjusted to 25 m^3/min and the flow rate can be monitored from the operator's cabin. To avoid blockages in the line, an extra water line is connected to the measurement line and in case of blockage, pressurized water is pumped into the line to clean it. To be on the safer side, additional gas sensors that can detect hydrocarbon gases were installed and integrated with the TBM as given in Table 14.2.

According to the hydrocarbon levels, if the gas concentration in the tunnel is between 10% and 20%, sensors give visual and audial warnings (blue light). If the gas concentration is between 20% and 40%, sensors that are integrated with the TBM and PLC stop the excavation and the personnel is evacuated from the tunnel. The tunnel is ventilated until the hazardous gasses are diluted below the threshold limit (orange light). Above 40%, the power is cut off and personnel are evacuated at once (red light). Periodical gas measurements were also carried out with portable gas detectors at various points of the TBM and throughout the tunnel.

The airflow was measured in different points of TBM by anemometers. It was detected that there were some areas without airflow reaches. It was presumed that

TABLE 14.2
Hydrocarbon Sensors Integrated to TBM

Channel No	Position of the Sensors
CH1	Under the screw conveyor discharge gate
CH2	Above the manlock door
CH3	Segment assembly area
CH4	Screw conveyor discharge gate
CH5	Segment feeder
CH6	Hydraulic motors
CH7	Next to chiller motor
CH8	TBM belt discharge

Source: Yazici et al. (2019).

these areas could be some sources for the accumulations of hazardous gases. Due to this reason, the exhaust type of ventilation was abandoned and replaced by blowing type ventilation. Since the contaminated area was close to the shaft, safety engineers decided that any additional ventilation was not required. Exhaust ventilation wasn't chosen since the toxic gas could spread along the tunnel and the air circulation wouldn't be enough to dilute hazardous gas inside the tunnel and the TBM. In case unsafe limits of hazardous gases were reached, the air velocity would increase.

Excavation works were carried out with the minimum staff possible. Aside from the key crew, no one was allowed inside the tunnel without fire-protective clothes or without the permission of the shift engineer. All activities liable to flame risks were prohibited during TBM advance. Emergency plans were developed and escape roads were kept continuously free. Considering unexpected system defects, calibration errors, or possible damage to gas sensors, manual gas detectors were also used. The excavation in the gasoline-contaminated area was successfully completed. This work was a typical example of how the situation was tackled in such contaminated ground by gasoline. If hazardous gas was detected, the problem was solved with an extra amount of air provided to the tunnel (Yazici et al. 2019).

14.3 CRITICAL SPACES IN A TBM FOR A FIRE ACCIDENT

The typical designs with diagrams of TBM are illustrated in Figures 14.4 and 14.5 to briefly introduce the inherent fire risks of TBM during operation. Fire protection

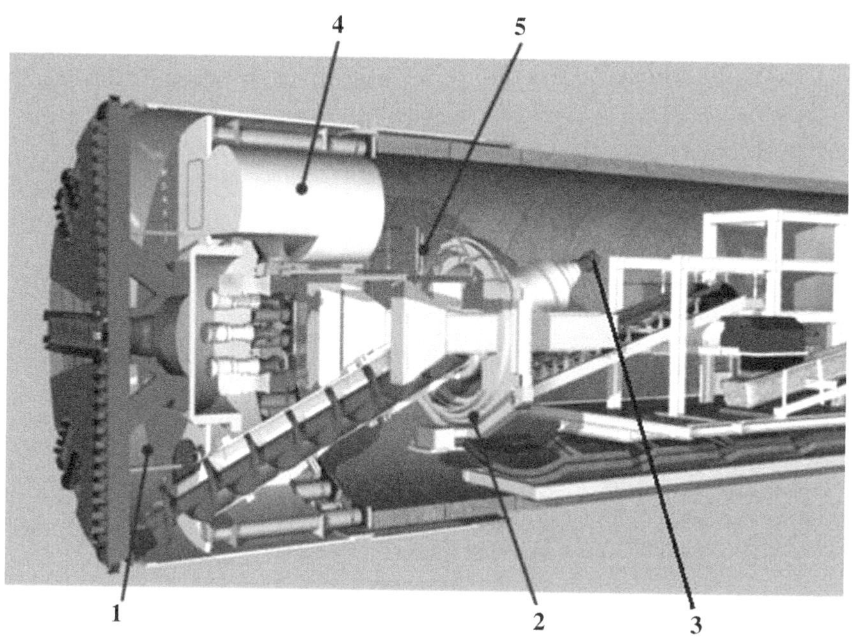

FIGURE 14.4 A typical design diagram of EPB-TBM. (BS EN 16191:2014.)

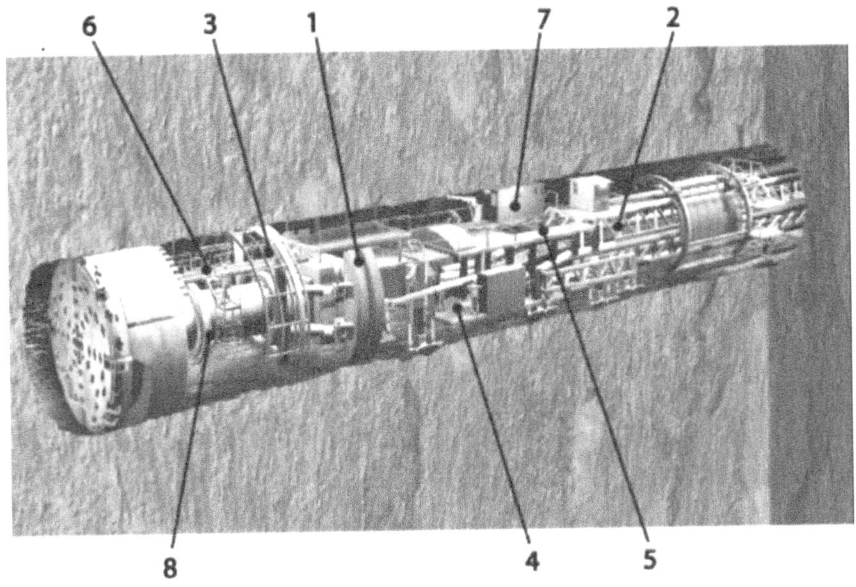

FIGURE 14.5   A typical design diagram of a gripper TBM. (BS EN 16191:2014.)

measures or the provisions of fire service installation (FSI) to be considered on TBM, adapted by the authors and originally published by Blennemann et al. (2005), are given in Table 14.3. It is a summary of recommendations for the planning and production of tunnel structures with regard to safety-related events compiled by DAUB (Deutscher Ausschuss für unterirdisches Bauen e. V.) [German Tunnelling Committee (ITA-AITES) Working Group]. We believe that these figures and tables should be considered altogether in any TBM project in order to design an efficient firefighting program.

The critical points in an EPB-TBM with inherent fire risks are: (1) excavation chamber, (2) lining erection equipment, (3) discharge area of screw conveyor, (4) air compressor area, and (5) working areas. The other critical points bearing fire risk are as defined in Figure 14.5. Utmost care should be taken in these areas for fire protection measures.

The critical areas in a gripper TBM with inherent fire risks are as follows: (1) gripping system, (2) towed back-up system, (3) erection device, (4) control station, (5) walkways, (6) working area, (7) main control station, and (8) rock bolting device. Utmost care should be taken in the areas defined in Figures 14.4 and 14.5 for fire protection measures and the provision of FSI. We believe that the means of access for firefighters to handle incidents at the TBM or coal mining site is crucial at the incipient stage of fire. However, this is beyond the expertise of the authors and the readers are advised to consult the fire and rescue service, operational guidance on incidents in tunnels and underground structures (CFRA 2012), and the code of practice for the provision of means of access for firefighting and rescue purposes published by Building Authority, Hong Hong (2004) to be better adequate with this topic.

TABLE 14.3

Fire Protection Measures or the Provisions of Fire Service Installation (FSI) To Be Considered on TBM

Fire Protection Measures	(A) Escape to Distance Safe Area (<400 m)	(B) Escape to Distance Safe Area (400–1,000 m	(C) Escape to Distance Safe Area (>1,000 m)
Communication system	One telephone	Two independent systems (phone/radio)	Like category (B)
Automatic or manual Extinguishing system	On transformers	On transformers or the other main sources of danger	Like category (B)
Portable extinguishers minimum 6 kg*	Distributed along the tunneling machinery	FPA	Like category (A)
Extinguishing water supply	Via existing service water pipes	Like category (A)	Like category (A)
Water curtains	–	On escape route and on TBM	Like (B), at $\varphi < 3.8$ m every 250 m
Automatic fire alarm systems	–	Fire alarm according to EN 12336	Like (B)
Rescue containers	–	–	Installation on trailers
Refuge chamber where shown necessary by the tunnel project risk assessment*	–	The tunnel crew plus 2 persons, minimum volume of 1.5 m³/person	Like (B)
Automatic sprinkler system	–	–	At $\varphi < 3.8$ m throughout The machinery area
Separate lighting with emergency power supply	–	–	At $\varphi < 3.8$ m with fire protected cables
Self-rescue units	With a 30-min hold time per person in the machinery area	Like (A)	At $\varphi < 3.8$ m self-rescue units storage every 1,000 m
Fire-protecting blankets	To be carried	Like (A)	Like (A)
Stocking of consumables	At the boring unit limited to one day	Like (A)	Like (A)
Storage of rescue equipment*	On all towed back-up equipment	Like (A)	Like (A)
Position finding aid	System providing information on the number of people in Tunnel	Like (A)	Like (A), additional measures for highly complex projects

TABLE 14.3 (Continued)
Fire Protection Measures or the Provisions of Fire Service Installation (FSI) To Be Considered on TBM

Fire Protection Measures	(A) Escape to Distance Safe Area (<400 m)	(B) Escape to Distance Safe Area (400–1,000 m	(C) Escape to Distance Safe Area (>1,000 m)
Emergency information*	A schema showing the escape routes, etc. close to back-up	Like (A)	Like (A)
Emergency crosscuts or shafts*			
Fire officer for the construction site			For full duration of construction site
Methane measurement*	10 cm at the top of TBM and next to the cutterhead	Like (A)	Like (A)

Sources: Blennemann et al. (2005) and Bilgin et al. (2021).

Notes: *Items added by the authors based on published standards as given in Table 14.4.
According to standard NFPA 130 (USA) in metro tunnels emergency escape shafts are necessary if the distance between two stations is greater than 762 m. For fire safety in twin tunnels, crosscuts are necessary for every 244 m.

14.4   MITIGATION MEASURES AGAINST FIRE OR METHANE EXPLOSION

Fire protection measures or the provisions of FSI including the standards used in different counties are summarized in Tables 14.3 and 14.4. These two tables will serve as a basis for mitigation measures against fire or gas explosion during TBM drives.

14.5   DIFFERENT STANDARDS RELATED TO THE SAFETY OF TBM, FIRE PREVENTION, AND PROTECTION

Breaking world records should never be tried on gassy ground as they did in the Los Angeles Tunnel (Proctor 2002). High gas emission limits, tunnel advance rates, and mining experience can be used in order to predict the methane inflow into a tunnel excavated through the gassy ground as explained by Rodriguez and Lombardia (2010). Experienced mining engineers should be employed as tunnel safety engineers since they have more experience than any engineers in difficult ground conditions. Gas measuring devices should be calibrated within acceptable periods. The explosion that occurred in Higashimurayama District in Japan is a typical example of this event (Kitajima 2010a). After detecting continuous gas emissions, all the electrical equipment should be replaced with flameproof equipment. This was not done in a

TABLE 14.4
Different Standards Related to the Safety of TBM, Tunnel Fire Prevention, and Protection Against Fire

Australia AS 2294.1-1997	Earth-moving machinery – protective structures
Australia AS 4825-2011	Tunnel fire safety
Austria ONORM B 2203-2:2005	Underground works – works contract – Part 2: continuous driving (TBM tunneling)
Austria ONORM EN 815:2009	Safety of unshielded tunnel boring machines and rodless shaft boring machines.
Bosnia & Herzegovina RVS-2001	Guidelines and regulations for road design
British BS 6164 fire-fighting facilities, evacuation and rescue facilities, fire detection alarm facility, ventilation and exhaust system, evacuation signs	Code of practice for safety in tunneling in the construction industry
British BS EN 13478:2001+A1:2008	Safety of machinery. Fire prevention and protection
British BS-EN-16191:2014	Tunneling machinery safety requirements
British BS 6164:2019	Soft ground and hard rock tunneling, pipe and box jacking
China DB43/729-2012	Road tunnel fire technical specification
China TB10063-2016	Railway Engineering Design Fire Code
China GB/T 33668-2017	Metro Safety Evacuation Code
European Union (EU) Directive 2004/54/EC	Minimum safety requirements for tunnels in the trans-European road network
European UNE EN 12111:2014	Tunneling machines, roadheaders, continuous miners, and impact rippers – safety requirements
European EN12336	Tunneling machines, shield machines, auger boring machines, lining erection machines – safety requirements
European EN1710	Equipment and components intended for use in potentially explosive atmospheres in underground mines
European EN12110 BS-EN-12111:2014	Tunneling machines, air locks, safety requirements
German RABT-2003	Guidelines for the equipment and operation of road tunnels
Turkey TS EN 815/April 1999	Turkish Standard for TBM working safety and fire
USA NFPA 502-2017	Standard for road tunnels, bridges, and other limited access highways
USA NFPA 130-2017. USA	Standard for Fixed Guideway Transit and Passenger Rail Systems
USA OSHA regulations, 29 CFR 1926.800	Fire prevention and protection in tunneling including underground machinery

Source: Bilgin et al. (2021).

water tunnel at the Zhinvalskaya hydro-power plant in Georgia, causing an explosion resulting in the death of all the crew. Remote sensing detectors should also be installed within the tunnel (Vlasov et al. 2001). Methane can dissolve in water and it may accumulate in different voids associated with tunnels, causing severe explosions as happened in the Abbeystead valve house. If water ingress is continuous in a gassy tunnel, methane emissions should be checked out until complete waterproofing of the tunnel is realized (Pearson et al. 1989; Lockyer and Howcroft 1997). Gas sensors should be installed just below the crown, not far away as they did not in Huyuki in Japan. Methane is lighter than air and accumulates at the crown of a tunnel (Kitajima 2010b). Accurate gas monitoring systems should be installed inside the TBM cutterhead, shield, and segment erector with an automatic TBM shutdown system. Preventive techniques such as grouting, pre-drainage, foam injection, and sealed lining should be used for gassy water inflow. Gas emissions may increase directly with the water ingress rate, so rapid dewatering systems should be established, as in Zagros Tunnel in Iran (Shahriar et al. 2009).

Different standards related to the safety of TBM, fire prevention, and protection are tabulated in Table 14.4 by order of the countries using these standards. It is strictly advisable for TBM users and tunnel constructors to follow these standards to avoid any disputes about safety issues. However, the comparison of these standards is beyond the scope of this paper. For those who are interested, it is advisable to consult a paper published by Li et al. (2018) which makes a comparison of the length of the tunnel. The main criticism of the comparative study was made separately of fire prevention in road tunnels, rail tunnels, and subway tunnels, emphasizing on fire-fighting facilities, evacuation and rescue facilities, fire detection alarm facilities, ventilation, and exhaust systems, and evacuation signs.

14.6 CONCLUDING REMARKS

Tunnel safety has increased significantly since previous tunnel fires in current tunnels. Much of the research studies carried out in the past on tunnel fires are related to completed tunnels, and there are only a few works on tunnel fires during construction or repair. Therefore, this chapter is aimed at tunnel fires that occurred during the construction process. The causes and examples of fires mainly due to hydraulic oil, belt conveyors, methane explosions, natural gas deposits, and ground contaminated by gasoline are treated in detail. Critical areas in a TBM for a fire accident are classified and shown in two illustrated TBMs. The points emerging from case studies, mitigation measures, and key points in fire management are discussed for fire or methane explosion. Different standards used in different countries related to the safety of TBM, fire prevention, and protection against fires are discussed in detail.

REFERENCES

Apte, V.B., 2006. *Flammability Testing of Materials Used in Construction, Transport and Mining*, 2nd edition. Woodhead Publishing, Abington, p. 464.

Bandini, A., Cormio, C., Berry, P., Battisti, M., Lisardi, A., Bernardini, P., Urso, M., 2019. Innovative solutions for safety against firedamp explosions in small section EPB-TBM tunnelling, in *Tunnels and Underground Cities: Engineering and Innovation meet Archaeology, Architecture and Art*, D. Peila, G. Viggiani & T. Celestino (eds), Taylor & Francis Group, London.

Beard, A., Carvel, R., 2011. *Handbook of Tunnel Fire Safety*, 2nd edition. ICE Publishing, London.

Bickel, J.O., Kuesel, T.R., King, E.H., 1996. *Tunnel Engineering Handbook*, 2nd edition. Kluwer Academic Publishers, Norwell, MA, p. 528.

Bilgin, N., Balcı, C., Aslanbaş, A, 2021. Case studies leading to the management of tunnel fire risks during TBM drives in an old coalfield, *Tunnelling and Underground Space Technology*, 112 (2021), pp. 103902.

Bilgin, N., Copur, H. and Balci, C., 2016. *TBM Excavation in Difficult Ground Conditions. Case Studies from Turkey*. Ernst & Sohn, Berlin.

Blennemann, F., Girnau, G., Grossmann, H., Pütz, R., Schreyer, J., 2005. *Fire Protection in Vehicles and Tunnels for Public Transport*, Alba-Verlag Düsseldorf, p. 504.

Building authority, Hong Hong, 2004. Code of practice for the provision of means of access for firefighting and rescue purposes, p. 28. www.bd.gov.hk/doc/en/resources/codes-and-ref erences/code-and-design-manuals/fs2011/fs2011_full.pdf

BS EN 16191:2014. (British Standards) Tunnelling Machinery – Safety Requirements.

Cafaro, E., Bertola, V., 2009. Fires in tunnels: Learning from disasters, *XXVII UIT Congress*, p. 9. https://techxplore.com/news/2020-03-tunnel-safety-minutes

CFRA (chief fire and rescue adviser) 2012. *Fire and Rescue Service, Operational Guidance Incidents in Tunnels and Underground Structures*, p. 198. https://assets.publishing.serv ice.gov.uk/government/uploads/system/uploads/attachm   ent_data/file/5917/2112377. pdf

Copur, H., Cinar, M., Okten, G., Bilgin, N., 2012. A case study on the methane explosion in the excavation chamber of an EPB-TBM and lessons learned including some recent accidents. *Tunneling and Underground Space Technology*, 27 (1), pp. 159–167.

D'Angelis, R., Maffucci, M., Giacomin, G., Secondulfo, Cichello, F., 2019. Pavoncelli Bis water tunnel: Tunnel boring machine selection and safety standards for excavating in presence of methane, in *Tunnels and Underground Cities: Engineering and Innovation meet Archaeology, Architecture and Art*, D. Peila, G. Viggiani & T. Celestino (eds), Taylor & Francis Group, London.

Doyle, B.R., 2001. *Hazardous Gases Underground: Applications to Tunnel Engineering*. Marcel Dekker, New York.

Francart, W.J., 2006. Reducing belt entry fires in underground coal mines, in 11th U.S./North *American Mine Ventilation Symposium*, J.M. Mutmansky & R.V. Ramani (eds), Taylor & Francis Group, London p. 7.

Hansen, R., 2009. Literature survey – fire and smoke spread in underground mines. *Studies in Sustainable Technology*, 2009, p. 2, Mälardalen University, p. 72.

Inal, M.E., Inal, N., 2015. Investigation on TBM performance in Silvan Tunnel depending on ground conditions. *Tunel, Journal of Turkish Tunneling Society*, Issue: March–April, pp. 54–62. (In Turkish).

Ingason., H. 2008. State of the art of tunnel fire research. *Fire safety Science-Proceedings of the ninth International Symposium*, University of Karlsruhe 21–26 September, pp. 33–48.

Ingason, H., Lönnermark, A., Frantzich, H., Maria Kumm, M., 2010. *Fire incidents during con-struction work of tunnels*. Fire Technology SP Technical Research Institute of Sweden, Report 2010:83, SP. ISSN 0284-5172.

Kitajima, M., 2010a. *Methane Gas Explosion Hazard During Construction of Headrace Tunnel for Agriculture*. www.shippai.org/fkd/en/cfen/CD1000099.html (accessed on 05.03.10).

Kitajima, M., 2010b. *Methane Gas Explosion Hazard of an Earth Pressure Type Shield Tunnel.* www.shippai.org/fkd/en/cfen/CD1000098.html. (accessed on 05.03.10).

Li, G., Han, W., Zhao, L., 2018. A Comparative Study of Fire Prevention Technical Standards for Tunnel Engineering at Home and Abroad. Atlantis Press, Tianjin.

Lockyer, J.W., Howcroft, A., 1997. The Abbeystead explosion disaster. *Annals of Burns and Fire Disasters*, 10, September 1–4.

Lönnermark, A., 2005. *On the characteristics of fires in tunnels*, Ph.D. Thesis Department of Fire Safety Engineering. Lund University.

Lönnermark, A., Hugosson, J., Ingason, H., 2010. *Fire incidents during construction work of tunnels – Model scale experiments,* SP Technical Research Institute of Sweden, SP Report 2010:86, p. 113.

Mitchel, D., Murphy, E.M., Smith, A.F., Polack, P.S., 1967. *Fire hazard of conveyor belts,* US Bureau of Mines RI 7053, p. 1.

Pearson, C.F.C., Edwards, J.S., Durucan, S., 1989. Methane occurrences in the Carsington Aqueduct tunnel project-a case study, *in Proceedings of the Rapid Excavation and Tunneling Conference*, pp. 176–195.

Proctor, R.J., 2002. The San Fernando tunnel explosion. *Engineering Geology*, 67, 1–3.

Rodriguez, R., Lombardia, C.R., 2010. Analysis of methane emissions in a tunnel excavated through carboniferous strata based on underground coal mining experience. *Tunneling Underground Space Technology*, 25, pp. 456–468.

Shahriar, K., Rostami, J., Hamidi, J.K., 2009. TBM tunneling and analysis of high gas emission accident in Zagros long tunnel, in *ITA World Tunnel Congress (WTC 2009)*, Budapest-Hungary, pp. 171–172.

Smith, A.C., Thimons, E.D., 2010. A summary of U.S. mine fire research. *SME Annual Meeting and Exhibit,* February 28–March 3, Phoenix, Arizona. Littleton, CO: Society for Mining, Metallurgy, and Exploration, Inc., pp.1–15.

Totten, G.E., De Negri, V.J., 2011. *Second Edition Handbook of Hydraulic Fluid Technology,* CRC Press, Taylor and Francis Group, p. 982.

Vlasov, S.N., Makovsky, L.V., Merkin, V.E., 2001. *Accidents in Transportation and Subway Tunnels*. Elex-KM Publications, Moscow, p. 198.

Yazici, H.A., Okkerman, M.B., Budak, C., 2019. Excavation through contaminated soil with EPB-TBM, WTC 2019, in *Tunnels and Underground Cities: Engineering and Innovation Meet Archaeology, Architecture and Art*, D. Peila, G. Viggiani & T. Celestino (eds), Taylor & Francis Group, London.

Index

Taylor & Francis eBooks

www.taylorfrancis.com

A single destination for eBooks from Taylor & Francis with increased functionality and an improved user experience to meet the needs of our customers.

90,000+ eBooks of award-winning academic content in Humanities, Social Science, Science, Technology, Engineering, and Medical written by a global network of editors and authors.

TAYLOR & FRANCIS EBOOKS OFFERS:

A streamlined experience for our library customers

A single point of discovery for all of our eBook content

Improved search and discovery of content at both book and chapter level

REQUEST A FREE TRIAL
support@taylorfrancis.com